# Photodegradation and Photostabilization of Coatings

# Photodegradation and Photostabilization of Coatings

**S. Peter Pappas,** EDITOR
*North Dakota State University*

**F. H. Winslow,** EDITOR
*Bell Laboratories*

Based on a symposium
sponsored by the Division of
Organic Coatings
and Plastics Chemistry
at the 179th Meeting of the
American Chemical Society,
Houston, Texas,
March 26–28, 1980.

ACS SYMPOSIUM SERIES 151

AMERICAN CHEMICAL SOCIETY
WASHINGTON, D. C.     1981

Library of Congress CIP Data

Photodegradation and photostabilization of coatings.
  (ACS symposium series; 151 ISSN 0097-6156)

  Includes bibliographies and index.

  1. Plastics—Deterioration—Congresses.  2. Photochemistry—Congresses.  3. Stabilizing agents—Congresses.  4. Plastic coating—Congresses.
  I. Pappas, Socrates Peter, 1936-   . II. Winslow,
Field Howard, 1916-   . III. American Chemical Society. Division of Organic Coatings and Plastics Chemistry. IV. Series: American Chemical Society. ACS
symposium series; 151.

TP1122.P48          668.4'9              81-467
ISBN  0-8412-0611-2                     AACR1
ASCMC 8        151        1–308          1981

# ACS Symposium Series

## M. Joan Comstock, *Series Editor*

# FOREWORD

The ACS SYMPOSIUM SERIES was founded in 1974 to provide a medium for publishing symposia quickly in book form. The format of the Series parallels that of the continuing ADVANCES IN CHEMISTRY SERIES except that in order to save time the papers are not typeset but are reproduced as they are submitted by the authors in camera-ready form. Papers are reviewed under the supervision of the Editors with the assistance of the Series Advisory Board and are selected to maintain the integrity of the symposia; however, verbatim reproductions of previously published papers are not accepted. Both reviews and reports of research are acceptable since symposia may embrace both types of presentation.

# CONTENTS

# PREFACE

Stabilization of organic coatings against photodegradation and photo-oxidation is a subject of obvious practical importance. The fundamental processes involved, including spectroscopy, photochemistry, energy transfer, and secondary reactions, have attracted scientists from diverse disciplines into this field.

This symposium deals primarily with current studies on the science and technology of coatings photostabilization. Related topics also are included with the expectation that exposure to a related field may serve to inspire new insights into one's specific discipline. An international scope is provided with contributors from Canada, France, Great Britain, Switzerland, West Germany, and the United States.

The topics include (1) mechanisms of action of stabilizers, including 2-(2'-hydroxyphenyl)benzotriazoles, 2-hydroxybenzophenones, and 2,2,6,6-tetramethylpiperidines; (2) synthesis of polymeric stabilizers; (3) interrelationships of energy transfer, photodegradation, and photoconduction in poly(N-vinylcarbazole); (4) photochemistry of model compounds; (5) photoactivity of $TiO_2$ pigments; (6) photodegradation and photostabilization of polymers and coatings, including bisphenol A-epichlorohydrin condensates, polyamides, polycarbonates, polyesters, polyethylene, polyvinyl chloride, and methacrylate varnishes for conserving museum objects; and (7) x-ray-induced degradation of photoresist polymer films.

The symposium was organized to provide an international forum for discussions of new concepts related to photodegradation, photooxidation, and photostabilization of organic coatings. It is hoped that the reader will share the enthusiasm of the participants.

S. Peter Pappas
Polymers and Coatings Department
North Dakota State University
Fargo, North Dakota 58105

F. H. Winslow
Bell Laboratories
Murray Hill, New Jersey 07974

November 25, 1980.

# Ultraviolet Stabilization of Polyamides

**Photophysical Studies of Ultraviolet Stabilizers, Particularly in the 2-Hydroxyphenyl Benzotriazole Class**

T. WERNER[1], G. WOESSNER, and H. E. A. KRAMER

Institut fuer Physikalische Chemie der Universitaet Stuttgart, Pfaffenwaldring 55, D-7000 Stuttgart 80, West Germany

The photodegradation of synthetic polymers can be considerably reduced upon addition of ultraviolet stabilizers. The UV stabilizers (preferably derivatives of o-hydroxy-benzophenone or of 2-(2'-hydroxy-5'-methylphenyl)benzotriazole (Tinuvin) transform the absorbed light energy into thermal energy thus preventing all sorts of photochemically initiated reactions. For review articles see the papers of Otterstedt (_1_), Heller and Blattmann (_2_,_3_), Kloepffer (_4_, _5_), Gysling (_6_) and Trozzolo (_19_).

In order to contribute to the elucidation of the mechanism of the UV-stabilization it seems reasonable to solve the following problems:

1° Is the UV-stabilization only due to the <u>screening effect</u> (or more precisely: light absorbing effect in a spectral region where the absorption spectra of polymer and UV-stabilizer overlap) of the UV-stabilizers and/or can it be enhanced by an <u>energy transfer</u> from the excited polymer to the stabilizer molecule?

2° From which <u>excited state</u> (<u>singlet</u> or <u>triplet</u>) of the polymer does the energy transfer take place?

3° Deactivation of the excited stabilizer molecule in its excited singlet (screening effect) and triplet state (energy transfer).

4° What is the origin of the rapid deactivation?

The system investigated was poly-(m-phenylene-

[1]Current address: Firma Bayer AG., FE-DPP-SF, D-5090 Leverkusen, FRG

*Scheme I.*

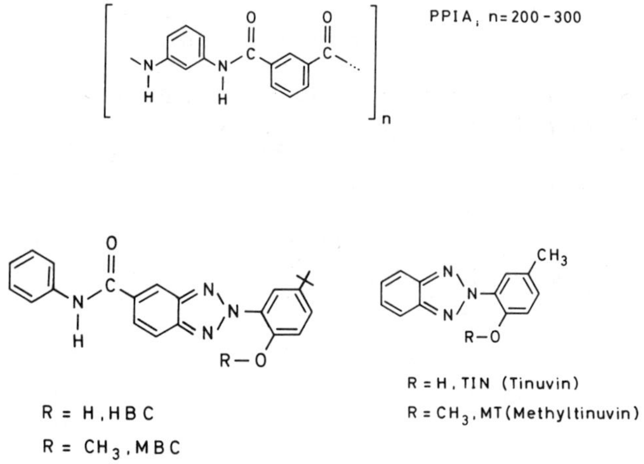

PPIA, n=200-300

R = H, HBC
R = CH₃, MBC

R = H, TIN (Tinuvin)
R = CH₃, MT (Methyltinuvin)

isophthalamide) (PPIA) as polymer donor and 2-(2'-
hydroxy-5'-tertbutylphenyl)-benzotriazole-carbonic-
acid-anilide-5    as stabilizer (acceptor) (HBC).
This stabilizer has been developed for PPIA by Küster
and Herlinger (7). The attached carbonic-acid-anilide
group should improve the compatibility of the stabilizer with the polyamide.

The original stabilizer (HBC) was modified: as
the rapid radiationless deactivation of the stabilizer
is (at least partly) due to the intramolecular hydrogen bond, the H-atom was substituted by a methyl group
(MBC). This "probe molecule" showed fluorescence and
phosphorescence and enabled us to demonstrate the
energy transfer to the stabilizer, simply by studying
its sensitized luminescence.

## Experimental Section

Fluorescence and phosphorescence spectra corrected for the instrumental sensitivity were measured with
a spectrometer described previously (8). Corrected excitation spectra were obtained with constant excitation intensity controlled by a rhodamine B quantum
counter. For phosphorescence polarization measurements
the apparatus was set up in an "In Line" arrangement
(9) and equipped with a Glan-Thomson polarizer and a
sheet polarizer (analyser) (10).

Polymer films were made from solution (8) and
spread on quartz windows. To remove oxygen the films
were kept for 2 hours at $10^{-5}$ Torr and then sealed off
in cylindrical quartz cuvettes.

From solutions, however, oxygen was removed by
bubbling with a stream of pure nitrogen (99.995 %)
for 45 min. Recrystallized TIN and its methoxy derivatives MT (and MBC) were gifts from Prof. Herlinger
and Dr. B. Kuester, (Institut fuer Chemiefasern,
Wissenschaftliche Institute an der Universitaet
Stuttgart).

For the investigation of triplet state properties
a laser flash photolysis apparatus was used. The excitation source was a Lambda Physik 1 M 50A nitrogen
laser which furnished pulses of 3.5 ns half-width and
2 mJ energy. The fluorescence decay times were measured with the phase fluorimeter developed by Hauser et
al. (11).

## Results and Discussion

### To question 1 (screening effect and/or energy transfer):

Fig. 1 shows the absorption spectra of polyamide films PPIA (curve A) and of the probe molecule MBC in films of ethylcellulose (curve B) (ethylcellulose films absorb at wavelengths shorter than 250 nm) and finally of polyamide films PPIA containing the probe molecule MBC (curve C). As the absorption spectra of polyamide and stabilizer overlap, the "screening effect" is evident. On the other hand, there is no wavelength at which the polyamide in film C can be selectively excited without exciting the probe molecule. Therefore, quantitative measurements of the luminescence (phosphorescence) of the probe molecule (MBC) upon excitation of polyamide films containing various concentration of MBC were undertaken in order to find out whether a sensitized phosphorescence (due to energy transfer) arises in addition to the directly excited phosphorescence.

In Fig. 2 the measured phosphorescence ($\lambda_{obs}$ = 525 nm) intensity $I_2$ of the probe molecule MBC in polyamide films is shown in relation to the MBC concentration (-190°C, oxygen free, $\lambda_{exc}$ = 289 nm). The curve $I_1 \times \gamma_\lambda$ represents the part of the MBC phosphorescence which is due to MBC molecules excited by direct absorption of light (screening effect!); thereby, $I_1$ is the relative phosphorescence quantum yield of MBC (measured in ethylcellulose films) while $\gamma_\lambda$ gives the fraction of the total incident light absorbed by MBC at the excitation wavelength $\lambda$, eq. 1, whereby the $\epsilon$ and c's denote the corresponding extinction coefficients and concentrations

$$\gamma_\lambda = \frac{\epsilon^\lambda_{MBC} \times c_{MBC}}{\epsilon^\lambda_{PPIA} \times c_{PPIA} + \epsilon^\lambda_{MBC} \times c_{MBC}} \qquad (1)$$

The difference between the total amount of phosphorescence ($I_2$) and the directly excited phosphorescence ($I_1 \times \gamma$) gives the sensitized phosphorescence ($I_{sens}$) the origin of which is the energy transfer from excited polyamide molecules to MBC. It can be seen that the sensitized phosphorescence is higher by a factor of 5 than the directly excited phosphorescence at the highest MBC concentration.

In order to corroborate the existence of energy transfer we studied the phosphorescence polarization. While the directly excited phosphorescence of MBC in

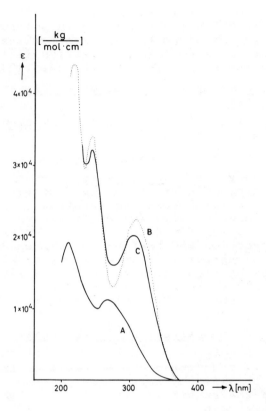

Angewandte Makromolekulare Chemie

*Figure 1.    Absorption spectra of: (A) PPIA (I), solid film, d ≈ 0.3 μm, $c_I$ = 4.2 mol/kg; (B) MBC (II), solvent: solid film of ethylcellulose, d ≈ 8 μm, $c_{II}$ = $10^{-2}$ mol/kg; (C) MBC (II), solvent: solid film of PPIA (I), d ≈ 0.3 μm, $c_{II}$ = 0.75 mol/kg. (8)*

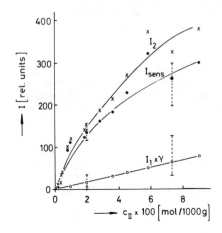

Angewandte Makromolekulare Chemie

*Figure 2.    Phosphorescence of MBC (II) in PPIA (I), in relation to MBC (II) (8).*

*$\vartheta$ = −190°C, excitation at 289 nm, emission observed at 525 nm, degassed films. ($I_2$) Measured total phosphorescence intensity of MBC (arbitrary units); ($I_1 \times \gamma$) component of the MBC phosphorescence due to direct absorption of MBC molecules; ($I_{sens}$) component of the MBC phosphorescence due to energy transfer.*

ethylcellulose films is positively polarized the polarization of MBC phosphorescence in polyamide films
is nearly zero as would be expected for sensitized
phosphorescence since the polarization is lost after
energy transfer steps (for details see (10)).

To question 2 (singlet-singlet or triplet-triplet
energy transfer):

      In Fig. 3 the total corrected emission spectrum
of MBC in ethylcellulose films is shown (curve A)
($\lambda_{exc}$ = 313 nm, -190$^{\circ}$C, oxygen free) while curve B
represents only the phosphorescence spectrum. The short
wavelength part of curve A is the fluorescence of MBC.
In curve C the total emission spectrum of polyamide
films containing 0.09 mol/kg MBC is shown. Correctly
speaking it is a difference emission spectrum. Since
no fluorescence of MBC is found (small amounts might
be due to the directly excited amount of MBC) we have
to conclude that the energy transfer is exclusively
a triplet-triplet energy transfer. (If it were a singlet-singlet transfer the excited singlet MBC would
emit fluorescence).

To question 3 (deactivation of the excited stabilizer
molecule in its excited singlet and triplet state):

      While we used the probe molecule to investigate
the energy transfer by sensitized phosphorescence we
now turn to the stabilizer itself (e.g. TIN with an
intramolecular hydrogen bond) to study its deactivation in the excited states.

## Absorption, emission and excitation spectra

      The absorption spectrum of TIN in methylcyclo-
hexane/isopentane at 150 K is represented by curve I,
Fig. 4. Curve III shows the absorption spectrum of MT
in hexane at 296 K. In unpolar solvents the intramolecular hydrogen bond of TIN is still intact (curve
I) whereas in polar solvents at least part of the TIN
molecules change this intramolecular into an intermolecular hydrogen bond to solvent molecules (curve
II). From this it must be concluded that the long-
wavelength absorption of curve I is due to the intramolecular hydrogen bond of TIN in unpolar solvents as
the intensity of this band is reduced in polar solvents (curve II) and disappears completely in the
spectrum of MT (without intramolecular hydrogen bond),
curve III. Curve IVa represents the fluorescence and
IVb the phosphorescence emission of both TIN and MT in

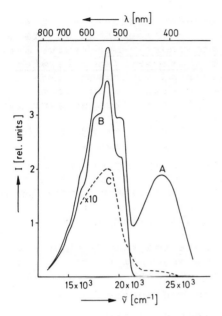

*Figure 3. Total emission and phosphorescence spectrum of MBC (II) at −190°C, corrected for the sensitivity of the detecting system (8).*

*(A) Total emission, matrix: ethylcellulose film, degassed, excitation at 313 nm, $c_{II}$ = 0.14 mol/kg; (B) phosphorescence, matrix: ethylcellulose film, degassed, excitation at 313 nm, $c_{II}$ = 0.14 mol/kg; (C) total emission, matrix: film of PPIA (1). The emission due solely to the polyamide film was subtracted, excitation at 302 nm, $c_{II}$ = 0.09 mol/kg.*

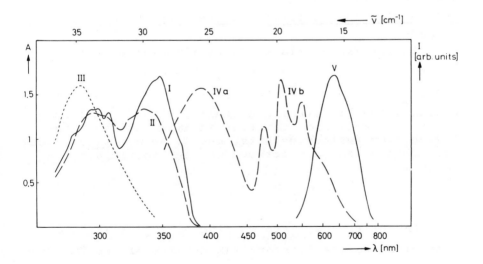

*Figure 4. Absorption and emission of TIN and MT in different solvents: (I) absorption of TIN in methylcyclohexane/isopentane at 150 K; (II) absorption of TIN in ethanol/methanol at 150 K; (III) absorption of MT in hexane at 296 K; (IV) (a) fluorescence and (b) phosphorescence of TIN or MT in 20:20:1 ethanol/ether/pyridine at 90 K; (V) fluorescence of TIN in methylcyclohexane/isopentane at 90 K.*

an ethanol/ether/pyridine mixture (20:20:1) at 90 K
(the emission intensity of MT is higher than of TIN).
     This fluorescence and phosphorescence (curve IVa
and IVb) originate only from those excited molecules
whose intramolecular hydrogen bond is broken. This is
proven by the phosphorescence excitation spectrum
where the long wavelength band (intramolecular hydrogen
bond) is lacking, Fig. 5, curve II.
     The very weak fluorescence of TIN at 638 nm (Fig.
4, curve V) with an unusually large Stokes shift of
7000 $cm^{-1}$ can only be observed in nonpolar solvents
(methylcyclohexane/isopentane, 90 K). It must be
attributed to TIN molecules with an intact intramolecular
hydrogen bond as its excitation spectrum (Fig. 5,
curve III) coincides with the absorption spectrum
(Fig. 5, curve I). The unusually large Stokes shift is
in agreement with our interpretation that a chemical
reaction takes place after the absorption(proton jumps
from O to N, $S_1 \rightarrow S_1'$, Fig. 6) and that the emission is
due to a new species (N-protonated species)$S_1' \rightarrow S_0'$.
     The reaction scheme presented in Fig. 6 is analo-
gous to Otterstedt's scheme (1) and contains in addition the triplet states.

Tautomerization equilibria in the ground and excited
states.

     In the ground state the equilibrium between $S_0$
(phenolic form N...H-O) and $S_0'$ (N-protonated form
N-H...O) is far on the side of $S_0$. The equilibrium
constant can be determined similarly to the method by
which the zwitterion constant in amino acids is obtained (for further details, see T. Werner (12)). The
equilibrium constant between the tautomeric derivatives is

$$K_Z = \frac{[\text{N-H...O}]}{[\text{N...H-O}]} \qquad (2)$$

     In the ground state we find $pK_Z(S_0) \leq 12$. Upon
excitation to the $S_1$ state the acidity of the OH group
increases tremendously as can be deduced from the
Förster cycle (13,14) and the proton jumps to the N
atom, the N-protonated form being now more stable than
the O-protonated form ($pK_Z(S_1) \leq -3 (\pm 3)$; $pK_Z(T_1) \leq 0$
($\pm 3$)). The $S_1'$ state is rapidly deactivated and fi-

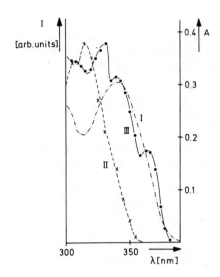

*Figure 5. (I) Absorption and (II) phosphorescence excitation spectra of TIN ($c_{TIN} = 2 \times 10^{-5}$M, d = 1 cm) in 20:20:1 ethanol/ether/pyridine at 90 K. Emission observed at 525 nm. (III) Fluorescence excitation spectrum of TIN (c = 2 × $10^{-5}$M) in methylcyclohexane/isopentane at 120 K. Emission observed at 615 nm.*

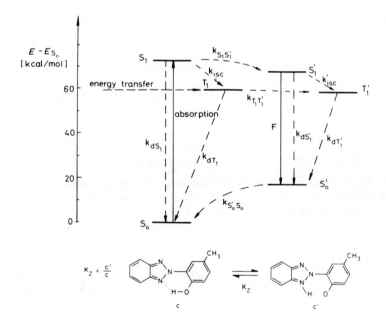

*Figure 6. Jablonski diagram for the excited-state proton transfer and energy dissipation in TIN: $k_{S_0'S_0}$, $k_{S_1S_1'}$, $k_{T_1T_1'}$ rate constants of proton-transfer processes in the ground state, first excited singlet state, and triplet state, respectively, and $k_d$ rate constants of radiationless deactivations and $k_{isc}$ rate constants of intersystem crossing.*

nally ends up in the $S_o'$ ground state of the N-protonated form which is unstable. The proton returns to the
O-atom and the molecule is restored to its original
state and is able to undergo further cycles by which
light energy is converted into thermal vibration energy (see also 17,18,19).

The change of $pK_Z$ by 15 units in going from $S_o$ to
$S_1$ is due to the fact that the acidity of the OH group
and the basicity of the $N_1$ atom simultaneously increase on excitation. Previous investigations (25)
show that the heterocyclic nitrogen basicity in quino-
xaline increases by 4-5 pK units upon excitation to
the first excited singlet state. One might expect similar behavior in the $N_1$ atom in this case. Additionally, it is well known (13,14)that the acidity of
naphtholic OH groups increase by roughly 7 pK units on
excitation. Therefore, the 15 unit change in $pK_Z$ observed here is consistent with what might be expected.
From the point of view of the proton in the intramolecular hydrogen bond the above $pK_Z$ changes exert a kind
of push pull effect. The charge densities calculated
by the CNDO/S method of Del Bene and Jaffé (15) for
$S_o$, $S_1$, $T_1$ are also in agreement with this interpretation. Therefore the change found in our case is
higher than in naphthols.

While the energy $S_o \rightarrow S_1$ is known from absorption measurements, the energy $S_1' \rightarrow S_o'$ is found from
the low temperature emission spectrum in nonpolar solvents (Fig. 4, curve V). The triplet energy of the
N...H-O form $(T_1 - S_o$; 60 kcal/mol) is determined at
room temperature in benzene solution by measuring the
quenching constant of TIN for several donors of different triplet energies. The triplet energy of N-H...O
form $(T_1' - S_o'$; 37 kcal/mol) cannot be measured and is
calculated semiempirically by the CNDO/S method of Del
Bene and Jaffé (15).

## Lifetime measurements of the excited states

Using phase fluorimetry the time dependence of
the $S_1' \rightarrow S_o'$ fluorescence $(\lambda_{max} = 638$ nm) can be described by two time constants, the shorter one being
mainly determined by $k_{S_1 S_1'}$ (rate constant of proton

transfer) while the longer one is governed by the
fluorescence decay time $\tau'$. For o-hydroxyphenylbenzo-
xazole both time constants could be determined, whereas for TIN only the longer one was found. (With higher
modulation frequencies we hope to extend the time
range of the phase fluorimeter up to 15 ps and to
measure the short time constant for TIN as well).
From this follows that in TIN $k_{S_1 S_1'} \leq 10^{11}$ s$^{-1}$.

This agrees quite well with the rate constants for
intramolecular proton transfer in 2,4-bis(dimethyl-
amino)-6-(2-hydroxy-5-methylphenyl)-5-triazine which
had been measured by Shizuka et al. (16) using laser
picosecond spectroscopy. The fluorescence decay constant $\tau'$ of $S_1'$ (TIN) was found to be 60 ± 20 ps. Because of the weak intensity all fluorescence lifetimes refer to the pure substance in crystalline form
at room temperature.

Using picosecond flash spectroscopy Gupta et al.
(24) reported for 2-hydroxyphenylbenzotriazole in
ethanol a short-lived transient (6 ps) followed by a
transient absorption whose lifetime is estimated to
be 600 ps. The authors assigned the short-lived transient to the "vertical singlet" while the long-lived
transient is presumably the "proton transferred species". These measurements of transient absorptions
with the picosecond flash method confirm our results
derived from the fluorescence emission using the phase
fluorimetric method.

It could be shown that the triplet ($T_1$) decay
time was < 20 ns in benzene at room temperature (see
Werner (12)). This means that both excited states of
the UV-stabilizer, namely singlet (produced by direct
light absorption, screening effect) and triplet (produced by energy transfer) are rapidly deactivated
whereby the electronic excitation energy is converted
to thermal vibration energy thus avoiding (or at least
diminishing) the photodegradation of the polymer and
of the UV-stabilizer itself.

Influence of deuteration upon the fluorescence $S_1' \rightarrow S_0'$
_____

The quantum yield of the fluorescence of the N-
protonated form ($S_1' \rightarrow S_0'$) is given by the following
expression

$$\Phi_F^H = \frac{k_{S_1S_1'}^H}{k_d^H + k_{S_1S_1'}^H} \times \frac{k_{F'}^H}{k_{F'}^H + k_{d'}^H} = \Phi_{Tr}^H \times \Phi_{F'}^H \qquad (3)$$

$$\Phi_F^H = \Phi_{Tr} \times \Phi_{F'}$$

$\Phi_{Tr}$ = Transfer yield; $\Phi_{F'}$ = Fluorescence yield whereby $k_{F'}^H$ is the emission rate constant for the transitions $S_1^1 \rightarrow S_o'$ in the undeuterated compound.

Upon deuteration (exchange of the H in the internal hydrogen bond for D) we get

$$\Phi_F^D = \frac{k_{S_1S_1'}^D}{k_d^D + k_{S_1S_1'}^D} \times \frac{k_{F'}^D}{k_{F'}^D + k_{d'}^D} = \Phi_{Tr}^D \times \Phi_{F'}^D \qquad (4)$$

We are able to measure the ratio $\Phi_F^H/\Phi_F^D$ and the fluorescence decay time of the deuterated and undeuterated compound $\tau'^D$ and $\tau'^H$, whereby $\tau'^D = (k_{d'}^D + k_{F'}^D)^{-1}$ and $\tau'^H = (k_{d'}^H + k_{F'}^H)^{-1}$. $\tau'^D$ is always longer than $\tau'^H$ in agreement with the well known deuterium effect that the rationless deactivation in deuterated compounds is slower ($k_{d'}^D < k_{d'}^H$) while $k_{F'}^D \approx k_{F'}^H$ remains unchanged. This leads to an increase of $\Phi_{F'}$ upon deuteration. From $\Phi_F^H/\Phi_F^D$ and $\tau'^H/\tau'^D$ the ratio of the transfer yields $\Phi_{Tr}^H/\Phi_{Tr}^D$ can be calculated if $k_d^D \ll k_{S_1S_1'}^D$ and $k_d^H \ll k_{S_1S_1'}^H$ hold. From $\Phi_{Tr}^H/\Phi_{Tr}^D$ some information about the change of $k_{S_1S_1'}$ upon deuteration can be obtained. For example in o-hydroxyphenylbenzoaxole $\Phi_{Tr}^D < \Phi_{Tr}^H$ is found. To obtain this result we used the fluorescence quantum yield of deuterated and undeuterated compound measured by Williams and Heller (26). Measurements of $\Phi_F^H/\Phi_F^D$, $\tau'^H$ at different temperatures are in progress which should enable us to decide whe-

ther the proton transfer in the excited state $(k_{S_1S_1'})$
is a tunneling process or needs an activation energy.

### To question 4 (Origin of the rapid deactivation)

The rapid deactivation of the excited singlet and
triplet state of TIN is due to the presence of the
__intramolecular hydrogen bond__, as shown by the long-
wavelength fluorescence of the N-protonated form at
$\lambda$ = 638 nm, $\tau'$ = 60 $\pm$ 20 ps (in crystalline form) and
the triplet decay time < 20 ns in a nonpolar solvent
(benzene). In polar solvents, however, the intramolecular hydrogen bond is broken and __intermolecular hydrogen bonds__ to solvent molecules are formed; as a consequence in rigid, polar solvents the fluorescence decay time of the O-protonated form ($\lambda$ = 390 nm) is
longer ($\tau$ = 2.5 $\pm$ 0.2 ns at 77 K) and the phosphorescence decay time is now in the range of
seconds (TIN, $\lambda$ = 508 nm, $\tau$ = 0.36 s in PPIA films and
$\tau$ = 0.58 s in ethylcellulose films at -180°C; MT, $\lambda$ =
525 nm, $\tau$ = 0.42 s in PPIA and $\tau$ = 0.56 s in ethylcellulose -180°C; HBC, $\lambda$=525 nm, $\tau$ = 0.26 s in PPIA
and $\tau$ = 0.58 s in ether/ethanol/pyridine at -180°C).
To understand this behavior  two points have to be
considered:

1° According to Otterstedt the energy difference in
   the N-protonated form $S_1' \rightarrow S_0'$ is lower than in the
   O-protonated form $S_1 \rightarrow S_0$ due to the proton transfer reaction, see Fig. 6. It is well known that the
   lower the amount of electronic energy which has to
   be converted into vibrational energy the faster
   this radiationless process (21). However, in addition to this argument, based only on __the amount__ of
   the electronic energy, there must be, as outlined
   above

2° a specific influence of the intramolecular hydrogen
   bond which accelerates the radiationless decay process.

For the rapid conversion of electronic energy to
vibrational energy a strong coupling between nuclear
and electronic motion is necessary (promoting and
accepting modes). Without going too much into details
we can say that strong coupling can be expected if
a) the potential curves in the ground and excited
   state are displaced
b) the vibrational frequency in ground and excited

state is different (distorted potential curves).
c) if the vibration is anharmonic.
   All three conditions seem to be fulfilled for the
intramolecular hydrogen bond.

$3^o$ According to Heller and Blattmann (3) the rotation
   of the hydroxyphenyl group around the central C-N bond
   may contribute to a rapid radiationless deactivation of
   the excited states. To understand this the
   items a) and b) of point $2^o$ with regard to a rota-
   tional vibration around the C-N bond can be offered
   as an explanation.
      The fluorescence of MBC in a solid matrix (ethyl-
cellulose films) is an order of magnitude higher than
in fluid solution (ethanol or tetrahydrofuran) (8).
A possible explanation might be as outlined above that
the excited singlet state is quenched by the rotation
of the 2-(2'-methoxy-5'-tertbutyl-phenyl)-group in analogy to the results obtained by Förster and Hoffmann
(22) with triphenylmethane dyes. In the compounds with
intramolecular hydrogen bonds (TIN, HBC), however, the
rotation and consequently the above mentioned quenching mechanism should be of minor import compared to
MBC. - In addition we have to remember that the fluorescence decay time of the $S_1'$ state (Fig. 6) amounts to
~ 100 ps. It is questionable whether the quenching by
rotation around the C-N bond can contribute to the
effective quenching of such a short-lived excited
state.

The polymer as solvent

      In this context the role of the polyamide as a
solvent should be considered. In Fig. 7 the absorption
spectra of TIN and of HBC in the polyamide PPIA are
shown. In the spectrum of TIN the intensity of the
long wavelength absorption band is diminished compared
with curves I and II of Fig. 4. From this we conclude
that part of the TIN molecules have formed intermolecular hydrogen bonds to the polyamide whereas for the
HBC molecules the intramolecular hydrogen bonds are
still intact (at least to a large extent). As the stabilizing efficiency is mainly due to the stabilizer in
form with the intramolecular hydrogen bond we expect a
higher stabilizing efficiency for HBC in PPIA than for
TIN in PPIA. That is exactly what has been found experimentally. (Due to the higher intensity of its long-
wavelength absorption the efficiency of the screening
effect of HBC is also greater than that of TIN).

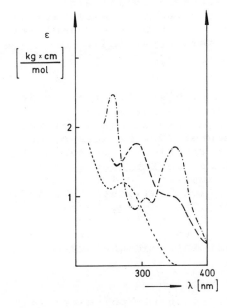

*Figure 7.   Absorption spectrum of HBC
(— · — ·), c = 0.7 mol/kg, and absorption of TIN (– – –), c = 1.7 mol/kg in
PPIA    film.    Absorption    spectrum    of
PPIA   film alone (· · ·).*

## Conclusions

In the system poly-(m-phenylene-isophthalamide) (PPIA) and o-hydroxyphenylbenzotriazoles studied in the present paper the UV-stabilization is due to the screening effect of the UV-stabilizer and to an energy transfer from the excited polymer triplet to the stabilizer triplet. Both the excited singlet state of the stabilizer (which is produced by the screening effect) and the triplet state (by energy transfer) are rapidly deactivated. The origin of this fast radiationless process is the intramolecular hydrogen bond.

Besides these "physical aspects" of the light protection of polymers there are some hints that UV-stabilizers of the o-hydroxyphenyl-benzotriazole type are able to scavenge radicals ("chemical aspects") the production of which could not be suppressed completely by the methods described in this paper (23).

## Acknowledgements

We thank Prof. Dr. H. Herlinger, Stuttgart, for having drawn our attention to this subject.

Thanks are due to Drs. G. Winter and U. Steiner for helpful discussions, to Prof. Dr. M. Hauser and Dr. H.-P. Haar for the phase fluorimetry measurements. For the chemical substances we are indebted to Prof. Dr. H. Herlinger and Dr. B. Kuester, Institut fuer Chemiefasern an der Universitaet Stuttgart, FRG, and Drs. H. Gysling and J. Rody, Ciba Geigy, Basel, Switzerland. The help of Dr. D.J. Miller, Stuttgart, and Dr. J.M. Menter, Atlanta, Georgia, USA, in translating the manuscript is gratefully acknowledged.

## Abstract

The photodegradation of synthetic polymers can be prohibited (or at least reduced) upon addition of UV-stabilizers. These compounds transform the absorbed light energy into thermal energy thus preventing all sorts of photochemically initiated reactions. Films of poly-(m-phenylene-isophthalamide) were used which contained appropriate amounts of UV-stabilizers of the 2-(2'-hydroxy-5'-methylphenyl)benzotriazole type or their 2'-methoxy derivatives. The following results were obtained:

1$^{\text{o}}$ The UV-stabilization effect is not only due to the screening effect of the UV-stabilizers but it is enhanced by energy transfer from the excited polymer to the stabilizer molecule.

$2^{\circ}$ The energy transfer occurs in the triplet manifold.

$3^{\circ}$ The stabilizer molecule in its excited singlet state (screening effect) and triplet state (energy transfer) is <u>rapidly</u> deactivated ($\tau(^1S)$) = 60 ± 20 ps, $\tau(^3T)$ < 20 ns for Tinuvin).

$4^{\circ}$ The internal hydrogen bond is the origin of the rapid deactivation.

## References

1. Otterstedt, J.A., <u>J.Chem.Phys.</u> 1973, 58, 5716
2. Heller, H.J. and Blattmann, H.R., <u>Pure and App-lied Chem.</u> 1972, 30, 145
3. Heller, H.J. and Blattmann, H.R., <u>Pure and App-lied Chem.</u> 1974, 36, 141
4. Kloepffer, W, <u>J.Polymer Sci.:</u> Symposium No. 57, 1976, 205
5. Kloepffer, W, Advances in Photochemistry, Vol. 10, 1977, 311
6. Gysling, H., <u>Kunststoffe</u> 1972, 62, 683
7. Kuester, B. and Herlinger, H., <u>Angew.Makromolek. Chem.</u> 1974, 40/41, 265
8. Werner, T.; Kramer, H.E.A.; Kuester, B.; Herlin-ger, H. <u>Angew.Makromolek.Chem.</u> 1976, 54, 15
9. Parker, C.A., Photoluminescence of Solutions, Elsevier Amsterdam, 1968, p. 229
10. Werner, T. and Kramer, H.E.A. <u>Europ.Polymer J.</u> 1977, 13, 501
11. Haar, H.-P.; Klein, U.K.A.; Hafner, F.W.; Hauser, M. <u>Chem.Phys.Letters</u> 1977, 49, 563
12. Werner, T., <u>J.Phys.Chem.</u> 1979, 83, 320
13. Foerster, Th., <u>Z.Elektrochem.</u> 1950, 54, 42
14. Weller, A. Progr.React.Kinet. 1961, 1, 189 G. Porter, editor, Pergamon Press, Oxford
15. Del Bene, J.A. and Jaffé, H.H. <u>J.Chem.Phys.</u> 1968, 49, 1221
16. Shizuka, H.; Matsui, K.; Hirata, Y.; Tanaka, I. <u>J.Phys.Chem.</u> 1976, 80, 2070; 1977, 81, 2243
17. Lamola, A.A. and Sharp, L.J. <u>J.Phys.Chem.</u> 1966, 70, 2634
18. Pitts, Jr., J.N.; Johnson, H.W.; Kuwana, T. <u>J.Phys.Chem.</u> 1962, 66, 2456
19. Trozzolo, A.M. in Polymer Stabilization, W.L. Hawkins, editor, Wiley-Interscience, New York 1972, p. 159
20. Turro, N.J. Molecular Photochemistry, W.A. Benja-min Inc., New York 1967, p. 69

21. Englman, R. and Jortner, J. Mol.Phys. 1970, 18, 145

22. Foerster, Th. and Hoffmann, G. Z.physik.Chem.N.F. 1971, 75, 63

23. Hodgeman, D.K.C., Journal of Polymer Science; Polymer Letters Edition 1978, 16, 161

24. Gupta, A.; Scott, G.W.; Kliger, D. Organic Coatings and Plastics Chemistry, Vol. 42, p. 490; preprints of papers presented by the Division of Organic Coatings and Plastics Chemistry at the American Chemical Society 179th National Meeting, Houston, Texas, March 23-28, 1980

25. Grabowska, A.; Herbich, J.; Kirkor-Kamińska, E.; Pakuła, B. J.Luminescence 1976, 11, 403

26. Williams, D.L.; Heller, A. J.Phys.Chem. 1970, 74, 4473

RECEIVED October 27, 1980.

# Photoenolization in Polymers

J. C. SCAIANO[1], J. P. BAYS, and M. V. ENCINAS

Division of Chemistry, National Research Council, Ottawa, Canada K1A 0R6
and Radiation Laboratory, University of Notre Dame, Notre Dame, IN 46556

A carbonyl chromophore in a macromolecule can participate in
a variety of photochemical processes that can have as end result
the degradation of the polymer via processes like the Norrish
Type I or Type II reaction, the triggering of a chain reaction
leading to peroxidation, the transfer of energy to another
chromophore or, it can also behave as an energy sink if a
suitable, non-degradative path, is available to the triplet state.
    The triplet state of carbonyl chromophores frequently shows a
high reactivity in hydrogen abstraction reactions (1). These
processes can take place intermolecularly (photoreduction) (1) or
intramolecularly, for example in the Norrish Type II process,
reaction 1 (2,3).

Frequently B will also undergo a back hydrogen transfer which
regenerates the parent ketone, as well as cyclization (in most
cases a minor reaction); as a result of this competition the
quantum yields of fragmentation are typically in the 0.1-0.5 range
in non-polar media. When the Norrish Type II process takes place
in a polymer it can result in the cleavage of the polymer backbone.
Poly(phenyl vinyl ketone) has frequently been used as a model
polymer in which this reaction is resonsible for its photodegradation, reaction 2.

[1]Current Address: National Research Council of Canada

0097-6156/81/0151-0019$05.00/0

$$\underset{\substack{\text{CO CO} \\ | \quad | \\ \text{Ph Ph}}}{\diagdown\!\!\!\diagup\diagdown\!\!\!\diagup\diagdown} \xrightarrow{h\nu} \underset{\substack{\cdot\text{COH CO} \\ | \quad | \\ \text{Ph Ph}}}{\diagdown\!\!\!\diagup\diagdown\!\!\!\diagup\diagdown} \longrightarrow \underset{\substack{\text{CO} \\ | \\ \text{Ph}}}{\diagdown\!\!\!\diagup\diagdown} + \underset{\substack{\text{CO} \\ | \\ \text{Ph}}}{\diagdown\!\!\!\diagup\diagup} \qquad (2)$$

Other reactions which excited carbonyl groups in polymers are
likely to be involved in include the Norrish Type I reaction
leading to the formation of free radicals, sensitized decomposition of peroxides or hydroperoxides and generation of singlet
oxygen, via direct quenching by oxygen, or indirectly if the
carbonyl triplet is quenched by a molecule which in turn generates
a long lived triplet.

Another common hydrogen transfer reaction of carbonyl triplets
is the photoenolization of the *o*-methylbenzoyl chromophore,
illustrated in reaction 3 for the *syn* conformer of *o*-methylaceto-
phenone (4). Reaction 3 can act as a very efficient energy sink,
and a number of properties of this group led us to believe that
this process could be used to reduce photodegradation; i.e. the
excellent absorption characteristics of the chromophore, the short
triplet lifetime and the fact that the disappearance of the
carbonyl triplet does not take place at the expense of the
formation of another excited state.

$$(3)$$

This paper reports a study of the photochemistry of polymers
and copolymers containing *o*-tolyl vinyl ketone units.

## Results and Discussion

The results presented in this report correspond to systems
where reactions 2 and 3 account for the decay of carbonyl triplets.
A series of copolymers of phenyl vinyl ketone and *o*-tolyl vinyl

ketone were examined. Deaerated solutions were irradiated with 366nm light and the progress of the photodegradation was monitored following the change of viscosity with time. Poly(phenyl vinyl ketone), PPVK, photodegrades with a quantum yield of 0.24, while poly($o$-tolyl vinyl ketone), PTVK, is photostable ($\Phi \leq 0.001$) even if both pathways (i.e. Norrish Type II and enolization) are in principle available to every $o$-methylbenzoyl chromophore, reaction 4.

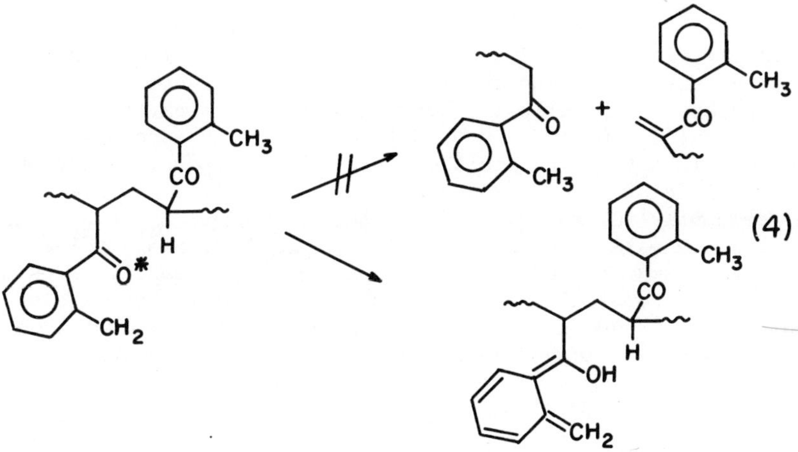

(4)

Perhaps the most interesting feature is exhibited by the copolymers, rather than the homopolymers. We find that the reduction in the quantum yield of photodegradation *always* exceeds the abundance of $o$-methylbenzoyl chromophores (5). Table I illustrates this effect.

TABLE I (6)

Effect of $o$-methylbenzoyl groups on the yield of photodegradation

| Polymer | Relative yield of degradation | % prevented |
|---------|-------------------------------|-------------|
| PPVK | 1.0 | 0 |
| CoPT(1)[a] | 0.81 | 19 |
| CoPT(3)[a] | 0.54 | 46 |
| CoPT(11)[a] | 0.08 | 92 |
| PTVK | 0 | 100 |

[a]The number in parenthesis indicates the relative abundance of $o$-methylbenzoyl chromophores.

The result can be interpreted in terms of efficient energy migration which allows the excitation to travel along the polymer, therefore favoring the faster decay process, i.e. photoenolization over the Type II process. We note that the triplet state energy of both chromophores is essentially the same. From an analysis of the dependence of the yields of degradation on the composition of the copolymer, we have estimated the average residence time of the energy in a given chromophore as ca. 30 ps (6). For comparison in the homopolymer, PPVK, the residence time was estimated as 1 ps (7). It should be noted that the efficiency of the o-methylbenzoyl group as an energy sink does not reflect a lower energy at that particular site (as is frequently the case in other systems), but rather a fast kinetic decay; in this sense the o-methylbenzoyl chromophore is a kinetically controlled energy sink.

Laser flash photolysis techniques offer the possibility of examining in detail the transient processes responsible for the photostabilizing effect discussed above. The triplet lifetimes are frequently too short, even for this technique; however, they can still be estimated using as a probe the quenching by 1-methyl-naphthalene, which leads to the formation of its easily detectable triplet. The optical absorbance due to the 1-methylnaphthalene triplet ($A_N$) produced as a result of energy transfer is related to the Stern-Volmer slope by equation 5, where N stands for

$$\frac{1}{A_N} = a + \frac{a}{k_q \tau_T [N]} \tag{5}$$

1-methylnaphthalene and 'a' is a constant that does not need to be evaluated explicitly. Figure 1 shows a typical plot according to equation 5. Similar plots for several copolymers lead to the data in Table II.

TABLE II (6)

| Polymer | $\Phi_{II}$ | $k_q \tau_T$/M-1 | $\tau_T$/M-1 | $\tau_B$/ns |
|---------|-------------|-------------------|--------------|-------------|
| PPVK | 0.24 | 130 | 55 | 65 |
| CoPT(1) | 0.194 | 110 | 46 | 75 |
| CoPT(3) | 0.130 | 75 | 31 | 160 |
| CoPT(11) | 0.020 | 38 | 16 | 250 |
| PTVK | 0 | 4 | 2 | 200 |
| CoMT(15) | – | 2.1 | 1 | 320 |

All values in benzene. $\Phi_{II}$ at 30°, all other values at 20°. The numbers in parenthesis indicate the percentage of o-tolyl groups.

Macromolecules

The biradicals produced in the hydrogen transfer reaction can be monitored directly using laser photolysis techniques (8).

The measurement of $\tau_B$ is straightforward when $\tau_T \ll \tau_B$. When this condition is not met, as is the case for the first two examples in Table II, the triplet lifetime can be shortened by addition of triplet quenchers, and the values of $\tau_B$ for PPVK and CoPT(1) have been obtained using this approach (6). Figure 2 shows a typical trace corresponding to the decay of the biradical from PTVK, as monitored at 415nm. The triplet state is in this case too short lived to be detectable; the residual absorbance observed after decay of the biradical is due to the enol.

Two types of biradicals can be produced in the copolymers, depending on whether the reaction occurs at a PVK or a TVK site. The change is reflected in the quantum yields, as well as in the lifetime for both, the triplet and the biradical, see Table II.

We have also examined the behavior of copolymers of *o*-tolyl vinyl ketone and methyl vinyl ketone (CoMT). In this case the light is absorbed exclusively at the aromatic carbonyl chromophore and the reaction proceeds from this site, while the methyl vinyl ketone moieties provide a relatively constant environment but prevent energy migration along the chain. The values of $\tau_B$ and $\tau_T$ in benzene have been included in Table II. These copolymers are also soluble in some polar solvents; for example, we have used a mixture of acetonitrile:acetone:methanol (30:30:40, referred to as AAM). This mixture is also a good solvent for the electron acceptor paraquat ($PQ^{++}$) which has been shown to be good biradical trap in a number of other systems (9).

The photolysis of CoMT(15) in the presence of $PQ^{++}$ leads to the occurrence of reaction 6.

$$\text{(6)}$$

The formation of the radical-cation, $PQ^{+\cdot}$ was monitored using laser photolysis techniques at its absorption maxima at 603nm. A study of the rates of $PQ^{+\cdot}$ formation at different $PQ^{++}$ concentrations led to $\underline{k_6} = 1.7 \times 10^9$ $M^{-1}s^{-1}$.[10] Despite the fact that this reaction is extremely fast, the rate of electron transfer for the macrobiradical is significantly slower than those for the same group in small molecules (8,11).

Finally, it should be mentioned that all our arguments and examples have centered on the *syn* conformer of the *o*-methylbenzoyl group. In small molecules the *syn-anti* conformational equilibrium is known to play an important role in the photochemistry of the

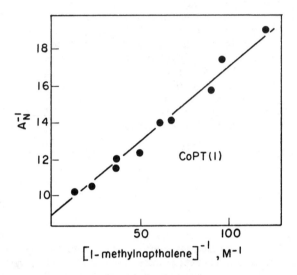

Figure 1.   Plot according to Equation 5 for CoPT(1).  Triplet 1-methylnaphthalene
was monitored at 420 nm.

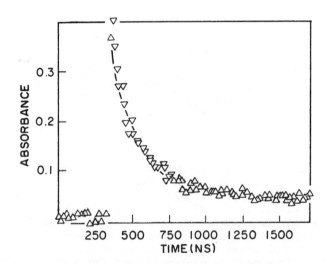

Figure 2.   Biradical decay for PTVK in benzene, and first-order fit of the data

molecules (4,11). In polymers, no such effect has been observed; apparently the *anti* conformer must play only a passive role in the process.

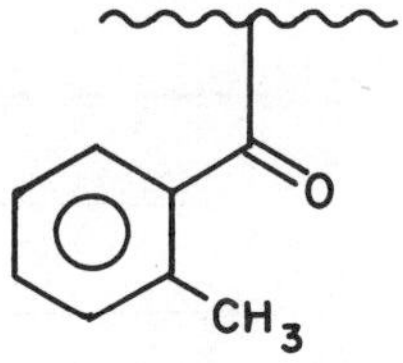

*syn* 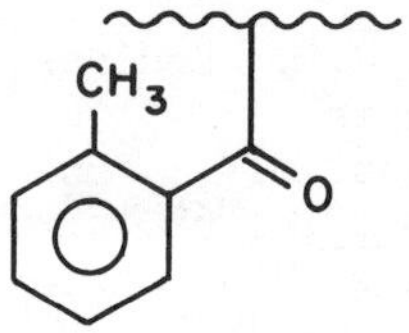                    *anti*

In conclusion, the *o*-methylbenzoyl group undergoes efficient radiationless decay via photoenolization involving the transfer of one of the benzilic hydrogens. The reaction occurs via the intermediacy of short lived triplets and biradicals. In copolymers containing aliphatic carbonyl groups the reaction centers at the *o*-methylbenzoyl chromophore as a result of its excellent absorption characteristics, while in the case of copolymers with phenyl vinyl ketone, energy migration 'funnels' the energy towards the photo-enolizable group leading to a considerable stabilization of the copolymer.

Experimental Section

Polymer and monomer preparation, purification and handling have been described in other reports from this laboratory (5,6,10).
Quantum yields of photodegradation were measured using an automatic viscosity timer modified so that the samples could be irradiated and deaerated *in situ*.
Polymer compositions were determined using $^{13}C$ NMR in the case of CoPT and UV spectroscopy of CoMT (5,10).
The laser photolysis experiments used the pulses (337.1nm, 8ns, 3mJ) from a Molectron UV-400 nitrogen laser for excitation. Details on the detection and processing of the data have been reported elsewhere (12).

Acknowledgements

The support from the National Research Council of Canada and the U.S. Department of Energy is gratefully acknowledged. This report has been issued as NRCC-17848 and NDRL-2060.

Literature Cited

1. Scaiano, J.C.; J. Photochem, 1973/74, 2, 81.
2. Wagner, P.J.; Acc. Chem. Res., 1971, 4, 168.

3.  Scaiano, J.C.; Lissi, E.A.; Encinas, M.V.; Rev. Chem. Intermed., 1978, 2, 139.
4.  Wagner, P.J.; Pure Appl. Chem., 1977, 49, 259.
5.  Bays, J.P.; Encinas, M.V.; Scaiano, J.C.; Macromolecules, 1979, 12, 348.
6.  Bays, J.P.; Encinas, M.V.; Scaiano, J.C.; Macromolecules, in press.
7.  Encinas, M.V.; Funabashi, K.; Scaiano, J.C.; Macromolecules, 1979, 12, 1167.
8.  See e.g. Haag, R.; Wirz, J.; Wagner, P.J.; Helv. Chim. Acta., 1977, 60, 2595.
9.  Small, Jr. R.D.; Scaiano, J.C.; J. Phys. Chem., 1977, 81, 828,2126. ibid, 1978, 82, 2662. J. Am. Chem. Soc., 1977, 99, 7713.
10. Bays, J.P.; Encinas, M.V.; Scaiano, J.C.; Polymer, 1980, 21, 283.
11. Das, P.K.; Encinas, M.V.; Small, Jr. R.D.; Scaiano, J.C.; J. Am. Chem. Soc., 1979, 101, 6965.
12. Encinas, M.V.; Scaiano, J.C.; J. Am. Chem. Soc., 1979, 101, 2146.

RECEIVED October 15, 1980.

# Mechanisms of Photodegradation of Ultraviolet Stabilizers and Stabilized Polymers

AMITAVA GUPTA—Energy & Materials Research Section, Jet Propulsion Laboratory, California Institute of Technology, Pasadena CA 91103

GARY W. SCOTT—Department of Chemistry, University of California—Riverside, Riverside, CA 92521

DAVID KLIGER—Division of Natural Sciences, University of California—Santa Cruz, Santa Cruz, CA 98064

Transparent, durable acrylic films which are opaque in the ultraviolet (300-380 nm) may be prepared by incorporating UV screening agents in a photo-stable acrylic such as PMMA (1,2). These UV screening acrylic films may then be used to protect polymers which commonly degrade when placed outdoors, e.g., polypropylene. However, UV screening agents which are blended in may be lost by evaporation or may be leached out. Moreover, blended in UV screening additives tend to aggregate in spots within the film and therefore cannot provide full protection. Durability of films containing UV screening agents pendant on the polymer backbone is determined by the stability of the additive itself, and the stability of the copolymer in which electronic energy transfer from the chromophore to the acrylic chain could lead to initiation of photodegradation. Both types of photo-processes leading to consumption of the UV screening agent involve interaction of the excited state with another molecule, e.g., oxygen or a neighboring ester group and may be attenuated by shortening the lifetime of the reactive excited state. It is found that the lifetime of the reactive excited state or a reactive tautomer should not exceed $1 \times 10^{-9}$ sec for outdoor photo-stability of up to twenty years for the additive itself. The photostability of the copolymer depends on the efficiency (or probability) of electronic energy transfer from the excited state of the UV screening agent to the repeating units of the base acrylic, e.g., methyl methacrylate. This energy transfer could take place through one or more of the following mechanisms: 1) electronic energy transfer which is normally absent if the carrier acrylic is PMMA, 2) hydrogen abstraction, leading to formation of a chain radical which may undergo chain scission, crosslinking or both processes depending on the radial reactivity and the mobility of local chain segments. Flash kinetic spectroscopy on 2-hydroxy benzophenone derivatives reported here and elsewhere (4,5) have allowed modeling of photodegradation of systems containing this class of UV screening agents as a comonomer. Quantum yield measurements reported here are in qualitative agreement with the predictive model proposed here.

0097-6156/81/0151-0027$05.00/0
© 1981 American Chemical Society

Preliminary data on another class of ultraviolet screening agents are also presented here. Excited state (transient) absorption and time resolved emission spectra have been recorded on 2(2'hydroxyphenyl) benzotriazole derivatives. These data correlate well with estimations of photostability of these UV screening agents blended with PMMA and cast into thin films. A series of copolymers of the benzotriazole class of UV screening agents with acrylics such as methyl methacrylate and n-butyl acrylate have been recently reported by O. Vogl and his group (14). These copolymers are currently under study in order to determine if copolymerization affects in any way the photostability or structure and decay kinetics of excited states of these UV screening chromophores.

## Hydroxybenzophenone Derivatives

There is some controversy regarding the decay mechanism of 2-hydroxy benzophenone. Scheme 1 summerizes all of the possible decay paths which involve proton transfer in the excited state and back transfer in the ground state.

*Scheme I.*

$$S_0(Ia)\text{-->}S_1(Ia)\text{-->}S_1(Ib)\text{-->}T_1(Ia)\text{-->}S_0(Ia) \qquad (1)$$

$$S_0(Ia)\text{-->}S_1(Ia)\text{-->}S_1(Ib)\text{-->}T_1(Ib)\text{-->}S_0(Ib)\text{-->}S_0(Ia) \qquad (2)$$

$$S_0(Ia)\text{-->}S_1(Ia)\text{-->}T_1(Ia)\text{-->}T_1(Ib)\text{-->}S_0(Ib)\text{-->}S_0(Ia) \qquad (3)$$

$$S_0(Ia)\text{-->}S_1(Ia)\text{-->}S_1(Ib)\text{-->}S_0(Ib)\text{-->}S_0(Ia) \qquad (4)$$

Here I(a) is the ketonic form and I(b) is the enol tautomer. Klopffer (6) has proposed that the main decay route involves the triplet manifold following proton transfer in the singlet state. Eisenthal (5) has recently reported kinetic data on this system in dichloromethane and ethanol at room temperature. He finds some residual transient absorption in ethanol lasting up to 10n sec, while there is no long-lived absorption in $CH_2Cl_2$. These results indicate that triplets which are formed in ethanol are relatively shortlived, and their decay is probably mediated by vibronic coupling to the O$\cdot\cdot$H—O band, while no detectable triplet formation takes place in $CH_2Cl_2$. This is consistent with the finding that the quenchable triplet yield is approximately 0·15 in ethanol while it is 0·03 in cyclohexane, further evidence for mechanisms (3) and (4) in Scheme 1. We have investigated photophysical and photochemical properties of 2-hydroxy benzophenone and a related system, poly (2-hydroxy, 3-allyl, 4,4'-dimethoxy benzophenone) co (mma), a copolymer containing approximately 0·5% (m/m) of the UV chromophore. Details of

synthesis and characterization of this copolymer have been
published (5,8). The molecular weight of the copolymer is about
46,000 ($\overline{M}_n$) and $\overline{M}_w/\overline{M}_n$ is about 1·89. This material may be
blended with PMMA to form a film from which the rate of loss of
chromophore is found to be negligible. This may be contrasted to
the rate of leaching of a commercially available high molecular
weight UV screening agent (Figure 1) from PMMA or PnBA.

### Photodegradation Studies on Copolymer Films

Rates of photodegradation of copolymer films were measured in
air using a filtered medium pressure Hg arc as the source of
radiation and o-nitrobenzaldehyde as the actinometer. Table 1
gives the dosage levels incident on these films.

Table I.    Radiation Input to the Copolymer Films. *(9)*

| Thickness x $10^{-3}$ | Time of Exposure,h | No. of Einsteins Absorbed Per $cm^2$ Per Second in Wavelength Range (nm) | | | |
|---|---|---|---|---|---|
| | | ⩽293 ($x10^{-10}$) | 293–300 ($x10^{-10}$) | 300–322 ($x10^{-8}$) | 323–386 ($x10^{-8}$) |
| 4 | 7 | 1.51 | 5.6 | 1.13 | 1.36 |
| 4 | 16 | 1.51 | 5.6 | 1.13 | 1.36 |
| 2 | 77 | 1.48 | 5.6 | 1.08 | 1.21 |
| 2 | 135 | 1.48 | 5.6 | 1.08 | 1.21 |
| 4 | 438 | 1.51 | 5.6 | 1.13 | 1.36 |

Macromolecules

Figure 2a and b show two typical types of spectral data. These
data indicate that the quantum yield of disappearance of the
chromophore is vanishingly small, i.e., less than 1 x $10^{-9}$. This
exceedingly low upper limit established in our experiments
correspond to an outdoor photostability of up to twenty years.
However, the copolymer undergoes photooxidation during the
exposure as shown in Figure 2b. This photooxidation is catalyzed
by the benzophenone chromophore because 1) all of the light is
absorbed by this chromophore and control films placed in dark at
room temperature for an equivalent period do not show similar
increase in absorbance at 3580 $cm^{-1}$, and 2) control PMMA films
exposed to the same radiation flux did not undergo any chemical
change. Changes in molecular weights of the films were
determined on an HPLC equipped with μ styragel columns and by
simultaneous refractive index and ultraviolet (285 nm) absorbance
monitoring. These results given in Table II show that the
overall average molecular weight decreases, while the average
molecular weight of the chains bearing chromophores,
approximately 67% of all chains, increases during the same
exposure period.
        The above results may be rationalized in terms of Scheme 2,
which proposes that the benzophenone chromophore catalyzes

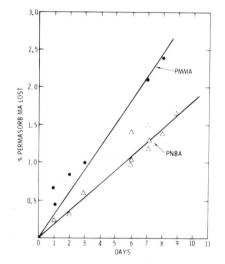

*Figure 1.   Rate of leaching of a UV sta-
bilizer from PMMA and PNBA when
those blends are immersed in water at
25°C*

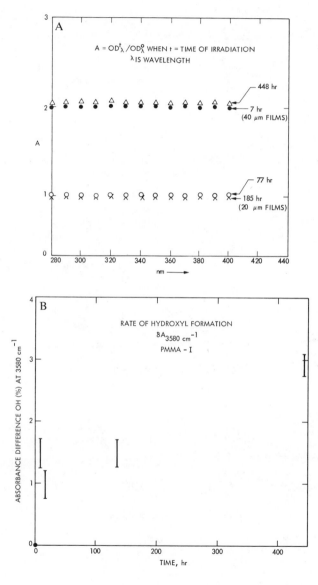

*Figure 2. Rates of photodegradation of the copolymer when exposed to a broad-band UV source (shortest wavelength 297 nm): (A) change in absorbance as a function of irradiation time in the wavelength range of 250–450 nm (spectrum measured on films); (B) change in absorbance at 3580 cm⁻¹ measured by FTIR spectroscopy (rate of hydroxyl formation)*

*Scheme II.*

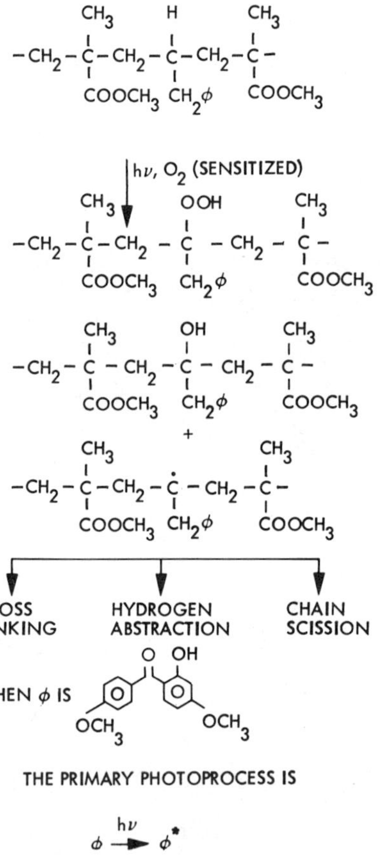

CROSS LINKING      HYDROGEN ABSTRACTION      CHAIN SCISSION

WHEN φ IS

THE PRIMARY PHOTOPROCESS IS

$$\phi \xrightarrow{h\nu} \phi^*$$

Table II. Molecular weight measurements on copolymer films undergoing irradiation. *(9)*

| Detection Mode | Time of Exposure (h) | $M_n$ | $M_w$ |
|---|---|---|---|
| RI | 0 | 40000 | 87000 |
| UV | 0 | 31000 | 69000 |
| RI | 448 | 32000 | 77000 |
| UV | 448 | 34000 | 79000 |

photooxidation of the tertiary hydrogen on the copolymer backbone. The mobility of the main chain radical bearing the methyl benzophenone moiety as the pendant group is expected to be higher than a PMMA chain radical because of the presence of the methylene group in the side chain and therefore this radical may undergo recombination and hydrogen abstraction as well as chain scission, much as higher polyalkyl methacrylates undergo simultaneous photo crosslinking and chain scission, while PMMA undergoes chain scission only[9]. The salient feature of the data is that the chain length of the chromophore bearing chains is increasing, so that the bleaching rate of the chromophore from a film consisting of PMMA blended with the copolymer will decrease with time. In any case, the quantum yields are exceedingly low, and the values reported in Table II are very close to the noise limit in these experiments.

Flash Spectroscopy

The apparatus used for picosecond flash spectroscopy on these systems has been described before(8,10). Figure 3a and b show typical transient absorption data obtained on 2-hydroxybenzophenone and the copolymer. Summary of these spectral data are given in Table 3. The transient observed at the shortest delay time (7ps) is the first excited singlet in all systems. The spectral data (at delay times > 50ps) permit placement of upper limits on triplet yields in $CH_2Cl_2$ for both 2-hydroxy benzophenone itself and the copolymerized chromophore. Nanosecond flash kinetic spectroscopy was also carried out on 2-hydroxy benzophenone and the copolymer (11). No transients could be detected in the nanosecond time scale, suggesting that the ground state enol [$S_1$ (Ib) in scheme 1] has a lifetime less than $1 \times 10^{-9}$ sec. These results strongly imply that processes (3) and (4) are responsible for the deactivation of singlet energy in these systems. A small, non zero triplet yield is postulated in the copolymer both to account for the photodegradation data and the transient spectral data. Triplet

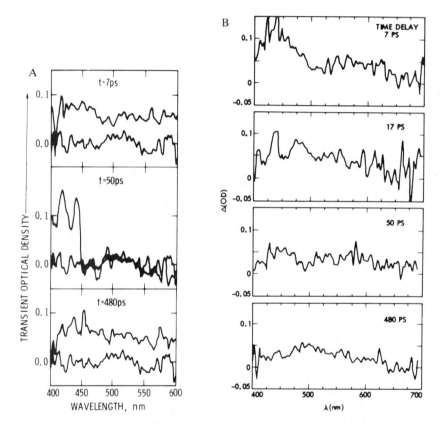

Figure 3. *Transient absorption spectroscopy on 2-hydroxybenzophenone and its derivative copolymer: (A) 2-hydroxybenzophenone in EtOH solution and 2-hydroxybenzophenone in CH₂Cl₂ solution; (B) copolymer in CH₂Cl₂ solution*

Table III. Transient Spectral data on 2-hydroxy benzophenone, and
its copolymer.*(9)*

| Molecule | Solvent | Delay Time ps | Max nm | OD at nm |
|---|---|---|---|---|
| 2-OH benzophenone | $CH_2Cl_2$ | 7 | 435 | 0.1 |
| | | 20 | 450 | < .04 |
| | | 485 | --- | <0.03 |
| | ETOH | 7 | 435 | .08 |
| | | 17 | <400; 475 | >0.1; .09 |
| | | 50 | 420 | 0.13 |
| | | 480 | 450 | 0.07 |
| copolymer | $CH_2Cl_2$ | 7 | 435 | 0.1 |
| | | 20 | 460 | <0.06 |
| | | 485 | --- | <0.03 |

Macromolecules

yield is considerable in ethanol which may lead to the prediction
that this class of UV stabilizers will be less effective in
elastomeric systems containing aliphatic hydroxyl groups.

## 2(2'-hydroxyphenyl) Benzotriazole Derivatives.

These are potentially superior UV screening agents, since
they have higher extinction coefficients (300-385nm) than
2-hydroxybenzophenone derivatives. Werner et. al. (12) have
proposed process (4) in scheme 1 as the major deactivation
pathway for 2(2'-hydroxy,5-methylphenyl) benzotriazole (II).
Kramer (13) has carried out an extensive study on the lifetime
and emissive properties of the excited states in the 5-methyl
derivative as well as on a methoxy derivative, e.g.
2(2'-methoxy,5-methylphenyl) benzotriazole. He estimates the
rate of proton transfer in the excited singlet state to be > $10^{11}$
$sec^{-1}$ while the lifetime of the proton transferred singlet was
found to be 100 ps from phase fluorimetry measurements. It is
estimated that the proton is favored to be on oxygen by
approximately 16 kCal/mole in the ground state,[14] while a change
of 15 pKz basicity units is expected in the excited singlet
state.

### Fluorescence Studies

Nanosecond flash spectroscopic studies showed that II
fluoresces in $CH_2Cl_2$ and ethanol solutions at room temperature.
The emission has a lifetime less than 1.0 nsec in either solvent
and cannot be resolved from the laser profile. Quantum yield of

emission in $CH_2Cl_2$ is exceedingly small at room temperature, of the order of $2 \times 10^{-5}$. Emission intensity is found to be even lower in ethanol. The maximum in emission occurs at approximately 450 nm, significantly blue-shifted from the spectra reported by Kramer at 190K. An acetoxy derivative was also prepared in order to eliminate proton transfer processes: 2(2'-acetoxy,5-methylphenyl) benzotriazole. Fluorescence quantum yield of this derivative is 0.038 and the lifetime is 0.2-0.6 nsec as measured by nanosecond flash spectroscopy.

## Transient Absorption Measurements

Transient absorption and ground state bleaching experiments were carried out using picosecond flash spectroscopy. Fig. 4a and b show transient absorption data obtained in $CH_2Cl_2$ at various delay times. Our tentative interpretation of these observations follows the decay mechanism proposed by Werner. The spectrum at 7 ps delay is assigned to excitation of the vertical singlet, $S_n \leftarrow S_1$, since the absorption occurs at the same wavelength as found in the acetoxy derivative. After 20 ps proton transfer has occurred in II, while by 485 ps internal conversion to the proton transferred ground state has occured. In support of this assignment, the absorption spectrum at 485 ps delay is approximately the mirror image of fluorescence ($S1^0 \leftarrow S^1$) from the proton transferred excited state. Ground state bleaching data are still being evaluated and will be reported in the future.

## Photochemistry

Photochemical changes in both II and the acetoxy derivative have been monitored in fluid solution and incorporated in a polymer film. Fig. 5 shows the spectral changes accompanying photochemical transformation of the acetoxy derivative. Then changes may be interpreted in terms of scheme 3, which proposes a photochemical 1,3 acyl shift to form "in situ" an ultraviolet stabilizer chromophore which also has a carbonyl functionality. The new compound has a better UV screening range and possibly superior quenching properties. Photodegradation of the UV stabilizer II was studied in PMMA films by UV visible spectroscopy, FT-IR scpectroscopy and HPLC. Fig. 6 shows some of the experimental data obtained. There is no spectral change of the stabilizer itself even at the longest irradiation period which is in agreement with the observation that we cannot detect a transient in the nanosecond flash kinetic experiments. Irradiation with 257nm radiation causing degradation (photooxidation and chain scission) of PMMA was partially inhibited by the presence of II due to the screening effect. The acetoxy derivative did not have an inhibitory effect, which is significant, since it has a high extinction coefficient at this wavelength, and therefore competes effectively with PMMA itself

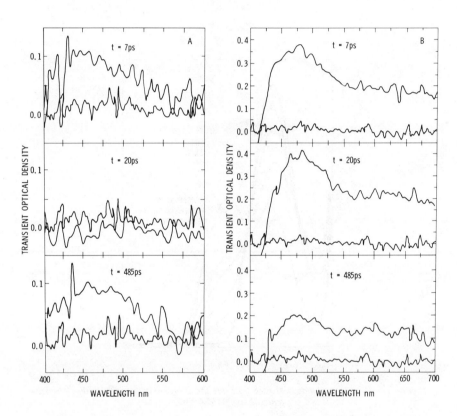

Figure 4.    *Transient absorption spectra of 2-hydroxyphenyl benzotriazole and its derivative in CH₂Cl₂*

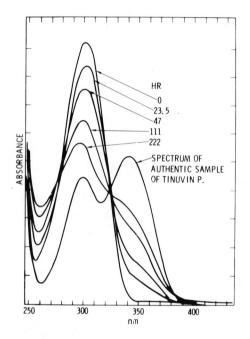

*Figure 5.   Photorearrangement rate data on the 2'-acetoxy-5-methylphenyl benzo-
triazole in PMMA films*

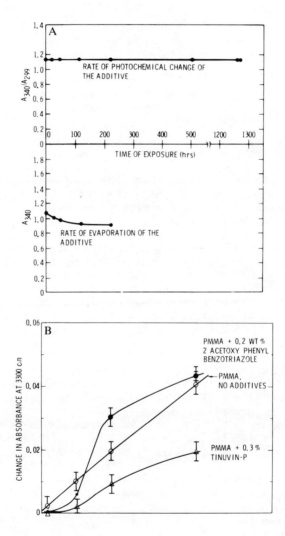

*Figure 6. Photodegradation rate data on Tinuvin-type UV stabilizers: (A) UV–visible spectrophotometric data; (B) FTIR data on PMMA films*

Scheme III.

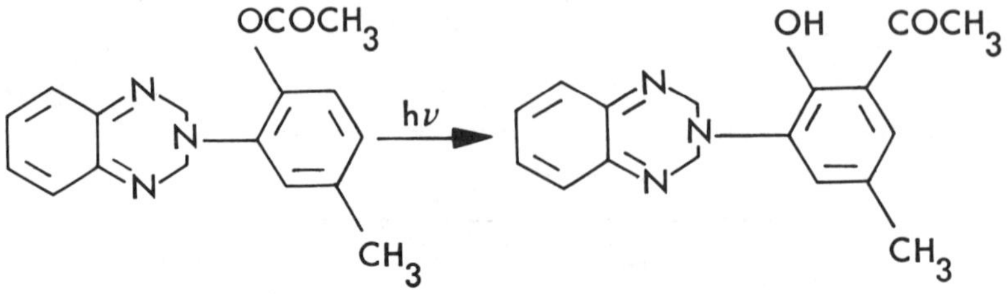

for radiation at 253.7nm. However as the acetoxy derivative is
converted to the photostabilizer it becomes an effective
inhibitor of photodegradation of PMMA. Inasmuch as the rate
controlling process in the mechanism of interaction of the
excited chromophore and the PMMA side chains is deactivation of
the excited state, the data in Fig. 6 shows that ultrafast
deactivation of excited states is essential if damage to the
carrier film is to be avoided. Preliminery experiments have been
carried out on copolymers of mma and vinyl benzotriazole
derivatives synthesized by Professor Vogl's group. Long
wavelength UV radiation [300 < λ (nm)<600] causes no detectable
damage to thin films for up to 1000 hours of irradiation, which
corresponds to 15 years of outdoor exposure. Measurement of
quantum yields of chemical changes are in progress.

Conclusions

Stabilization of degradable polymers through incorporation
of ultraviolet stabilizers is never a complete success, since
photodegradation may always take place at the surface where
protection is incomplete. Hence photodegradation gradients build
up in such systems. These problems are avoided by encapsulating
these polymers in clear ultraviolet screening films which are
themselves stable. The stability of the protected systems is
then dependent on the stability of the UV screening films.
Chemical incorporation of UV screening agents onto chain segments
of stable acrylics such as PMMA was attempted and demonstrated in
order to minimize physical loss of UV screening agents from these
films. It was then necessary to evaluate the photostability of
the copolymers, and also the ability of the immobilized UV
screening chromophore to rapidly deactivate electronic energy.
Quantum yield measurements and flash kinetic spectroscopic
techniques were used to gather data which demonstrate that the
copolymer films may be expected to perform their function for
long periods outdoors and that lack of mobility of the UV

screening agents does not affect their screening properties in the systems studied so far. Thus these preliminary results point to a new approach to development of protected polymers for outdoor use.

## Abstract

Quantum yields and lifetimes of emission (fluorescence) as well as other principal rates of deactivation have been measured on 2-hydroxy benzophenone and 2-hydroxyphenyl benzotriazole derivatives. Polymerizable UV screening agents have been prepared and copolymerized with acrylics in order to obtain transparent films containing nonfugitive UV screening agents. Preliminary results of studies of photodegradation on these copolymers are also reported here.

## Acknowledgements

This article describes one phase of research performed at the Jet Propulsion Laboratory and supported by the Department of Energy under an agreement with the National Aeronautics and Space Administration.

## Literature Cited

1. Trozzolo, A. M.; Polymer Stabilization, W. L. Hawkins ed., Wiley-Interscience, New York 1972, pp. 159.
2. Heller, H. J., and Blattman, H. R.; Pure Appl. Chem., 30, 145 (1972).
3. Huston, A. L., Merritt, C. D., Scott, G. W., and Gupta, A; Paper presented at the Topical Meeting on Picosecond Phenomena, Cape Cod, Mass., June 18, 1980.
4. Merritt, C., Scott, G. W., Gupta, A., and Yavrouian, A.; Chem. Phys. Lett., 69, 109 (1980).
5. Hon, S. Y., Hetherington, W. J., III., Koreschowski, G. M., and Eisenthal, K. B.; Ibid, 68, 282 (1979).
6. Klopffer, W.; J. Polym. Sci.; Symp. 57, 205 (1976).
7. Klopffer, W.; Adv. in Photochem., 10, 311 (1977).
8. Werner, T.; J. Phys. Chem. 83, 320 (1979).
9. Gupta, A., Yavrouian, A., Di Stefano, S., Merritt, C., and Scott, G. W.; Macromolecules, 13 (4) 821 (1980).
10. Gupta, A., Liang, R., Tsay, F., and Moacanin, J.; Macromolecules, In press.
11. Anderson, R. W., Jr., Damaschen, D. E. Scott, G. W., and Talley, L. D.; J. Chem. Phys., 71, 1134 (1978).
12. Gupta, A., Kliger, D., and Liang, R.; Unpublished Results.
13. Werner, T., and Kramer, H. E. A.; Europ. Polym. J., 13, 501 (1977).
14. Kramer, H. E. A.; Private Communication; Lecture at the Photodegradation and at the Photostabilization of Polymeric

      Films and Coatings Symposium, ACS National Meeting, Houston,
      March 1980.
15.   Tinell, D., Bailey, D, Pinazzi, C., and Vogl, O.,
      Macromolecules, 1978, 11, 312.


RECEIVED October 14, 1980.

4

# Preparation of Polymeric Ultraviolet Stabilizers

DAVID A. TIRRELLDAVID A. TIRRELL

Department of Chemistry, Carnegie–Mellon University, Pittsburgh, PA 15213Department of Chemistry, Carnegie–Mellon University, Pittsburgh, PA 15213

A consideration of conventional ultraviolet stabilizers used to prevent or retard the photooxidation of organic polymers, shows that overall effectiveness is often severely limited by poor long-term performance. This is not, in general, a result of photodecomposition of the stabilizer, but rather of physical loss of the stabilizer from the matrix polymer through diffusion, through extraction or through phase separation processes. The solubility of the stabilizer in the matrix is often less than the minimum effective concentration, leading to stabilizer migration, and exposure to solvents and/or high temperatures in processing or in use can accelerate stabilizer loss. These problems are particularly acute in the stabilization of thin films and coatings.

It is primarily the prospect of reduced mobility and volatility, with the expected improvement in long-term performance, which motivates the preparation of polymeric ultraviolet stabilizers. Work in this area was reviewed thoroughly by Bailey and Vogl in 1976 (1), and more recently by the author (2). The present paper describes recent synthetic work involving four classes of effective ultraviolet stabilizers: salicylate esters (I), 2-hydroxybenzophenones (II), α-cyano-β-phenyl-cinnamates (III) and hydroxyphenylbenzotriazoles (IV). In each

0097-6156/81/0151-0043$05.00/0
© 1981 American Chemical Society0097-6156/81/0151-0043$05.00/0
© 1981 American Chemical Society

case, the approach has involved the preparation of a modified
stabilizer which is in fact a substituted styrene, and two
classical styrene syntheses have been employed (Scheme I).  In
Scheme Ia, the aromatic ring of the stabilizer is acetylated

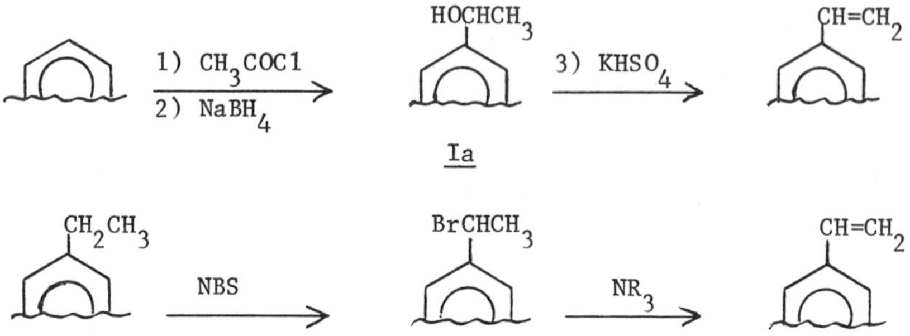

Scheme I

in a Friedel-Crafts reaction, and reduction to the benzylic
alcohol followed by dehydration produces the polymerizable
stabilizer derivative.  This route is particularly useful in
the synthesis of relatively volatile compounds (e.g. the
salicylates) since the dehydration is extremely rapid and the
product can be distilled immediately from the hot reaction flask.
   The sequence in Scheme Ib has been more generally useful.
The accessibility of ethyl-substituted stabilizer precursors
suggests benzylic bromination with N-bromosuccinimide followed
by base-catalyzed dehydrobromination. Tertiary amines were found
to be particularly effective in the syntheses described in this
paper.
   Radical polymerizations of vinyl-substituted ultraviolet
stabilizers were accomplished with azobisisobutyronitrile (AIBN)
as initiator, with careful exclusion of oxygen.  Copolymerization
was also readily achieved.  The following sections describe in
detail the preparation of polymeric ultraviolet stabilizers from
salicylate esters, 2-hydroxybenzophenones, $\alpha$-cyano-$\beta$-phenyl-
cinnamates and hydroxyphenylbenzotriazoles.

   Salicylate Esters.  The methyl esters of three isomeric
vinylsalicylic acids (the 3-, 4- and 5-vinyl compounds) have
been prepared, using the synthetic routes outlined above.
Methyl 5-vinylsalicylate was prepared as shown in Scheme II (3).
Friedel-Crafts acetylation of methyl salicylate gave the 5-acetyl
derivative in 80% yield.  Blocking of the phenol followed by
NaBH$_4$ reduction then provided methyl 5-(1-hydroxyethyl)acetyl-
salicylate in an overall yield of 70%.  The hydroxyethyl
compound was dehydrated over KHSO$_4$ at 225°C/0.2 mm, and the

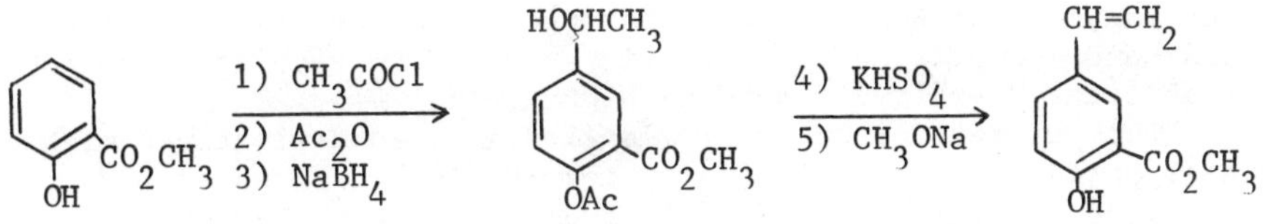

Scheme II

phenol freed by treatment with sodium methoxide in methanol. The overall yield of methyl 5-vinylsalicylate was 35%. Methyl 5-vinylsalicylate was polymerized and copolymerized with vinyl monomers, using AIBN as radical initiator, apparently without interference by the phenolic hydroxyl groups.

The preparative route to the 3- and 4-vinyl compounds is illustrated in Scheme III, for the preparation of methyl 3-vinylsalicylate (4).

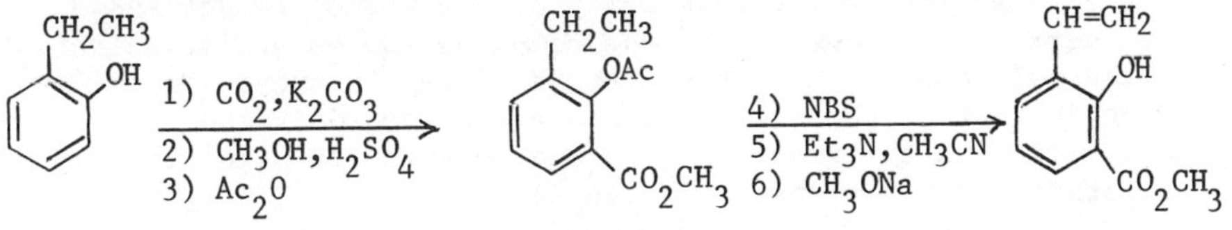

Scheme III

2-Ethylphenol was carbonated in the position ortho to the hydroxyl group by treatment with $CO_2$ under pressure, in the presence of anhydrous $K_2CO_3$ at 175°C. The yield of 3-ethylsalicylic acid was 72%. Esterification with methanol, acetylation with acetic anhydride, and benzylic bromination with N-bromosuccinimide afforded methyl 3-(1-bromoethyl)acetylsalicylate in an overall yield of 42%. NBS was superior to $Br_2$ in $CCl_4$ in the benzylic bromination; use of the latter reagent produced a mixture of the desired product plus methyl 3-(1-bromoethyl)-$\alpha$-bromoacetylsalicylate. Dehydrobromination with triethylamine in acetonitrile, followed by treatment with sodium methoxide in methanol, gave methyl 3-vinylsalicylate in 22% yield overall from 2-ethylphenol. Triethylamine in acetonitrile proved to be a convenient dehydrobrominating agent; the reaction was complete after five hours, and isolation of the product was facilitated by precipitation of the amine hydrobromide and by the volatility of the solvent. The yield of the dehydrobromination was 83% using this reagent, vs. 56% with tri-n-butylamine in dimethylacetamide.

A similar route--carbonation, bromination, dehydrobromination--produced methyl 4-vinylsalicylate in 28% yield from 3-ethylphenol (5).

Three alternative routes to 3-vinylsalicylic acid derivatives were also investigated briefly, without success: carbonation of

2-hydroxyacetophenone, Fries rearrangement of acetylsalicylic
acid, and formylation of salicylic acid. None gave the desired
3-substituted salicylic acid in high yield.

The polymerization of vinylsalicylic acid derivatives might
be expected to be complicated by termination or transfer
reactions involving the phenolic hydroxyl group. Our observations concerning this point seem to parallel those of Kato (6),
who studied the radical polymerization of hydroxystyrenes with
AIBN an initiator. The kinetics and mechanism of the
polymerizations of meta- and para-hydroxystyrenes were found to
be those of typical radical polymerization, while intramolecular
deactivation of the growing chain was postulated to account for
reduced molecular weight and anomalous kinetic behavior in the
polymerization of the ortho-hydroxy compound. In the vinylsalicylic acid derivatives prepared in this work, the hydroxyl
group is involved in strong hydrogen-bonding with the carbonyl
of the neighboring ester or acid function, and the effect of
this interaction on the reactivity of the phenol is not known
with certainty. However, if one compares the molecular weights
of the polymers of the methyl esters of 3-, 4- and 5-vinyl-
salicylic acids, the suggestion is again that only the hydroxyl
group ortho to the growing chain end contributes significantly
to termination or transfer (Table I).

## Table I

### Inherent Viscosities of Poly(Methyl Vinylsalicylates)

| Polymer | $\eta_{inh}$ |
| --- | --- |
| Poly(methyl 5-vinylsalicylate) | 2.46 d$\ell$/g (DMSO) |
| Poly(methyl 4-vinylsalicylate) | 2.61 d$\ell$/g (DMSO) |
| Poly(methyl 3-vinylsalicylate) | 0.16 d$\ell$/g (Benzene) |

The data in Table I are not directly comparable, since the
viscosity of the 3-isomer was determined in benzene while the
others were measured in DMSO. In addition, the first two
polymers were prepared in bulk polymerizations, while the
polymerization of methyl 3-vinylsalicylate was carried out with
the monomer diluted 1:1 with benzene. Thus no certain conclusion
can be drawn; the data are, however, an indication of possible
difficulty in radical polymerization of substituted styrenes
bearing a phenol ortho to the vinyl group.

2,4-Dihydroxy-4'-Vinylbenzophenone.  2,4-Dihydroxy-4'-
vinylbenzophenone was prepared in 30% yield from 4-ethylbenzoic
acid as shown in Scheme IV (7).

Acylation of resorcinol by 4-ethylbenzoic acid gave
2,4-dihydroxy-4'-ethylbenzophenone in 72% yield. Attempted
bromination of this compound with NBS yielded not the benzylic
bromide, but rather ring-brominated products. Since ring-

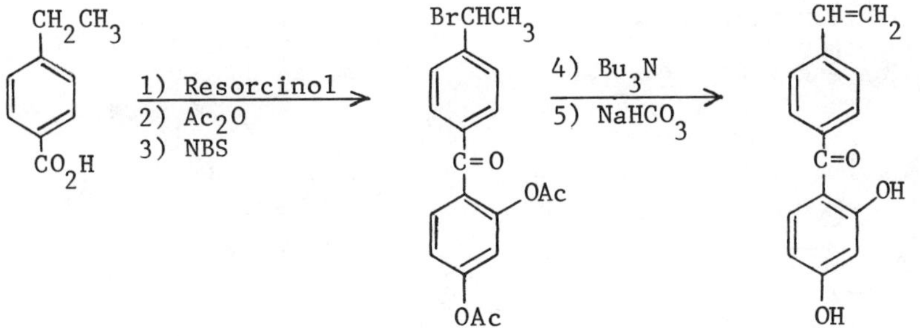

Scheme IV

bromination with NBS had previously been observed for aromatic
compounds with strongly electron-donating substituents (8), the
phenols were blocked by acetylation, and the bromination repeated.
A 75% yield of 2,4-diacetoxy-4'-(1-bromoethyl)benzophenone was
obtained. Dehydrobromination with tri-n-butylamine in DMAc,
followed by treatment with NaHCO$_3$ in aqueous methanol produced
the desired 2,4-dihydroxy-4'-vinylbenzophenone.

2,4-Dihydroxy-4'-vinylbenzophenone was converted to a
homopolymer of inherent viscosity 0.57 d$\ell$/g by polymerization
with AIBN in dimethylformamide. The UV spectrum of the polymer
showed the three absorption maxima characteristic of 2,4-
dihydroxybenzophenones (at 324, 292 and 248 nm), although the
extinction coefficient was depressed in comparison with the
4'-ethyl analogue.

The behavior of 2,4-dihydroxy-4'-vinylbenzophenone in
radical copolymerization with styrene was quite unexpected. The
presence of two phenolic hydroxyls again suggests possible
transfer or termination in radical polymerization, so we
examined carefully the effects of adding small amounts of 2,4-
dihydroxy-4'-vinylbenzophenone or 2,4-dihydroxy-4'-ethylbenzo-
phenone to a radical polymerization of styrene. The results
were dependent on the initiator used: with AIBN, the phenolic
compounds did not change the polymer molecular weight at all,
within experimental error, while with benzoyl peroxide, an
increase in molecular weight was observed. This was ascribed
to the differing reactivities of the initiator radicals, and it
was suggested that only the benzoyloxy radical from benzoyl
peroxide reacts by abstraction of the phenolic hydrogen atom.
The net result is a decrease in initiator efficiency and an
attendant increase in polymer molecular weight. In fact, all
of our work on radical polymerization of phenol-containing vinyl
monomers suggests that inhibition and transfer problems are at
most minor, if AIBN is used as initiator and oxygen is carefully
excluded from the reaction mixtures (9).

Ethyl 4-Vinyl-α-Cyano-β-Phenylcinnamate.  Ethyl 4-vinyl-α-
cyano-β-phenylcinnamate has been prepared recently by Sumida,
Yoshida and Vogl (10).  The route is shown in Scheme V.

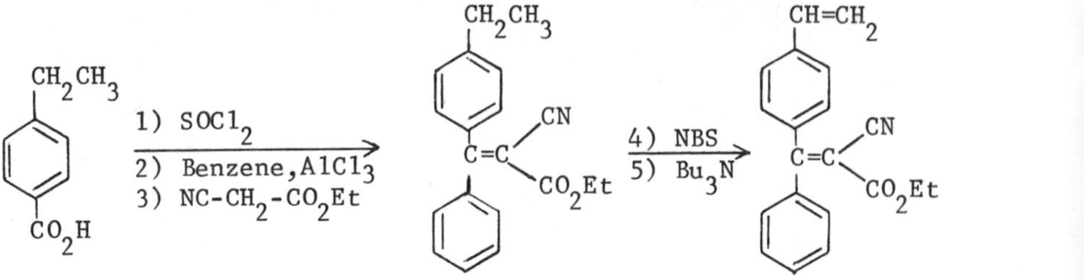

Scheme V

4-Ethylbenzoic acid was converted to the acid chloride, which
was treated with AlCl$_3$ and benzene to give 4-ethylbenzophenone
in 90% yield overall.  Condensation with ethyl cyanoacetate
afforded ethyl 4-ethyl-α-cyano-β-phenylcinnamate as an essentially
50/50 mixture of the Z- and E-isomers.  The yield of the condensation was highly sensitive to reaction conditions, and was
optimized at 75% with portionwise addition of the ammonium
acetate catalyst.  Bromination and dehydrobromination as
described earlier then completed the preparation.  The overall
yield of ethyl 4-vinyl-α-cyano-β-phenylcinnamate was 20%.

Homopolymerization of ethyl 4-vinyl-α-cyano-β-phenylcinnamate
with AIBN in benzene gave a soluble polymer of inherent viscosity
0.2 dℓ/g.  There was no evidence for involvement of the tetra-
substituted double bond in the polymerization.  Copolymerizations
with styrene and methyl methacrylate were also successful.

2-(5-Vinyl-2-hydroxyphenyl)benzotriazole.  Yoshida and Vogl
have recently prepared 2-(5-vinyl-2-hydroxyphenyl)benzotriazole
by the route shown in Scheme VI (11).

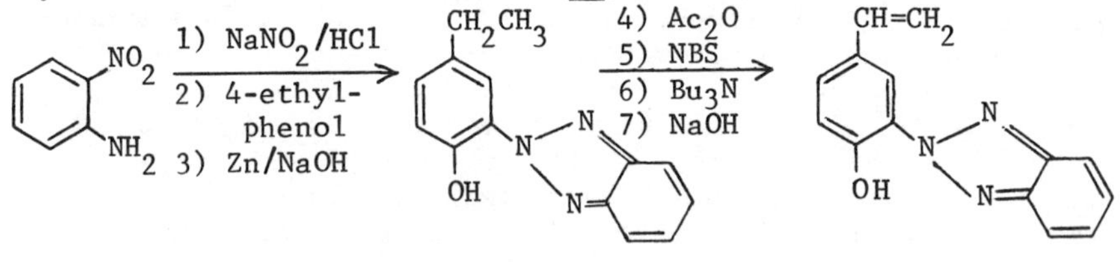

Scheme VI

The hydroxyphenylbenzotriazole structure was constructed by a
coupling of the diazonium salt of o-nitroaniline with 4-ethyl-
phenol, followed by reduction of the nitro-azobenzene to the
benzotriazole with zinc powder and NaOH.  After blocking of the
phenol by acetylation, bromination and dehydrobromination were
performed as described earlier, and treatment with aqueous NaOH

provided 2-(5-vinyl-2-hydroxyphenyl)-benzotriazole in an overall
yield of 20%. Homopolymerization and copolymerizations with
styrene and methyl methacrylate have been accomplished.

Conclusions

The incorporation of four different classes of important
ultraviolet stabilizers into high polymer chains has been
accomplished by synthesis of polymerizable, vinyl-substituted
stabilizer derivatives followed by radical polymerization.
Each of the derivatives may be regarded as a substituted
styrene, and classical styrene syntheses have been employed.
Radical polymerization of the phenolic monomers (salicylate
esters, 2-hydroxybenzophenones and hydroxyphenylbenzotriazoles)
proceeds normally with AIBN as initiator, at least when oxygen
is carefully excluded. It is expected that polymeric ultraviolet
stabilizers, perhaps in combination with conventional stabilizers,
will make an important contribution to photostabilization
technology.

Acknowledgement

The author would like to thank Professor Otto Vogl for
permission to read and use several manuscripts prior to
publication.

Literature Cited

1. Bailey, D. and Vogl, O.  J. Macromol. Sci., Rev. Macromol.
   Chem., 1976, 14(2), 267.

2. Tirrell, D.  Polymer News, in press.

3. Bailey, D.; Tirrell, D.; and Vogl, O.  J. Polym. Sci., Polym.
   Chem. Ed., 1976, 14, 2725.

4. Iwasaki, M.; Tirrell, D.; and Vogl, O.  J. Polym. Sci.,
   Polym. Chem. Ed., in press.

5. Tirrell, D. and Vogl, O.  Makromol. Chem., in press.

6. Kato, M.  J. Polym. Sci., A-1, 1969, 7, 2175.

7. Bailey, D.; Tirrell, D.; Pinazzi, C.; and Vogl, O.
   Macromolecules, 1978, 11(2), 312.

8. Horner, L. and Winkelman, E. H. in W. Foerst, Ed. "Newer
   Methods of Preparative Organic Chemistry," Vol. 3, Academic
   Press, 1964, p. 176.

9. Tirrell, D.; Ph.D. Thesis, University of Massachusetts, 1978.

10. Sumida, Y.; Yoshida, S.; and Vogl, O.  Polymer Preprints,
    1980, 21(1), 201.

11. Yoshida, S. and Vogl, O.  Polymer Preprints, 1980, 21(1), 203.

RECEIVED September 16, 1980.

# Polypropylene Photostabilization by Tetramethylpiperidine Species

D. J. CARLSSON, K. H. CHAN, and D. M. WILES

Division of Chemistry, National Research Council of Canada, Ottawa, Canada K1A 0R9

The photo-oxidative degradation of polypropylene (PPH) can be largely summarized by the reactions 1 to 4 (1,2). During the

$$PPOOH \xrightarrow{\text{h}\nu} PPO\cdot + \cdot OH \xrightarrow{PPH} \text{some } PP\cdot \qquad 1$$

$$PP\cdot + O_2 \xrightarrow{k_2} PPO_2\cdot \qquad 2$$

$$PPO_2\cdot + PPH \xrightarrow{k_p} PPOOH + PP\cdot \qquad 3$$

$$2\ PPO_2\cdot \xrightarrow{2k_t} \text{some non-radical products} \qquad 4$$

course of photo-oxidation, the near ultra violet (UV) photo-cleavage of the macro-hydroperoxide (PPOOH) produced by reaction 3 is a dominant source of free radicals (1). The dramatic deterioration in the mechanical properties (tensile strength, elongation at break) which occurs at relatively slight degrees of photo-oxidation in PPH (about one tert. C-H in a thousand) is associated with the β-scission of the macro-alkoxyl radicals produced either in the photo-initiation (reaction 1) or in the complex self-reaction of the tert.-peroxyl radicals (reaction 4) (2).

In principle, photo-stabilization can result from the prevention of any of the processes 1 to 3. Photo-initiation may be reduced by the addition of an inert UV absorbing additive, or quenching the excited states of the initiating species (PPOOH, carbonyl, etc) before dissociation to radicals can occur. The former process is ineffective at low additive levels (∿0.1w%) in thin articles (<50μm) and cannot protect front surfaces. Energy quenching of excited PPOOH groups has not been demonstrated. Although light replaces heat as an essential component of the initiation step, steps 1 to 4 are in fact identical to the processes involved in the classical thermal oxidations of alkanes.

0097-6156/81/0151-0051$05.00/0

Consequently conventional antioxidant mechanisms must be expected
to protect against photo-oxidation.  Thus hydroperoxide decomposition to inert molecular products will reduce the rate of photoinitiation and scavenging of any of the free radical species will
be beneficial, although the effectiveness of conventional antioxidants in photo-oxidations is limited by their own stability
and the photo-sensitizing propensity of their products (3).

The hindered secondary amines can be highly effective photo-
stabilizers for various polymers (4,5,6).  Various hindered amines
have been shown to retard oxidation, but most share the common
feature of being secondary or tertiary amines with the α-carbons
fully substituted.  The most widely exploited representatives of
this class are based on 2,2,6,6-tetramethylpiperidine either in
the form of relatively simple low molecular weight compounds, or
more recently as backbone or pendant groups on quite high molecular weight additives (4,5,6).  The more successful commercial hindered amines contain two or more piperidine groups per molecule.
Photo-protection by tetra-methylpiperidines (near UV transparent)
must result from the interruption of one or more of the reactions
1 to 3.  Relatively recent results from our own laboratories, and
in the open literature will be outlined in this context.

In order to simplify the experimental problems involved in
unravelling the mechanisms of UV protection by the piperidines,
we have concentrated on the use of the simpler monopiperidine
compounds.  Although our findings are relevant to the photo-
protection by the more complex multifunctional, commercial
additives, some major differences may exist, and will be
emphasized together with the very significant effects of the solid
state on photo-stabilization.

## Photoprotection by Simple Piperidines

### Radical Scavenging Processes

In photo-oxidizing PPH, piperidine stabilizers do not persist
beyond the relatively early stages of irradiation.  For example,
during irradiation in a xenon arc Weather-Ometer, $5 \times 10^{-3}$M of
4-oxo-2,2,6,6-tetramethylpiperidine in a 25μm PPH film is un-
detectable after only 100h, whereas accumulation of polymer
oxidation products is delayed until ∿400h (7).  Thus the parent
piperidine is apparently not essential for photo-stabilization,
although it can make a contribution to radical scavenging and
PPOOH decomposition in the early stages of irradiation.  For
example, peroxyl radicals will slowly attack the $>$NH group, with
the generation of the stable nitroxyl free radical (reaction 5,
$k_{NH} \sim 3 M^{-1} s^{-1}$ at $25°C$ in the liquid phase) (8).  An approximately

$$>NH + RO_2 \cdot \xrightarrow{k_{NH}} ROOH + >N \cdot \xrightarrow{O_2} >NO \cdot \qquad 5$$

quantitative yield of nitroxyl radicals is also generated very
slowly by a dark reaction when a pre-oxidized PPH film is exposed

to a piperidine solution ($\tau_{\frac{1}{2}} \sim 30$ days) presumably by the usual amine-induced, hydroperoxide decomposition mechanisms which can be summarised as reaction 6 (9).

$$\text{>NH} + \text{HOOR} \rightleftharpoons [\text{>NH} \text{---} \text{HOOR}]$$

$$\text{>NO}\cdot + \text{H}_2\text{O} + \text{ROH} \xleftarrow{\text{ROOH}} \text{>N-OH} + \text{HOR}$$

6

The nitroxide radical (from processes 5 and 6 and attack by other radicals on the parent piperidine) is found in photo-oxidizing PPH samples in concentrations of $\sim 1 \times 10^{-4}$M (initial piperidine level $5 \times 10^{-3}$M) up to the embrittlement point of the PPH film (7). Nitroxides are well known to scavenge carbon centered radicals (but not peroxyl radicals) in both polymers and liquid alkanes (reaction 7) (10, 8). In the liquid phase $k_7$ is

$$\text{>NO}\cdot + \text{PP}\cdot \xrightarrow{k_7} \text{>NOPP}$$

7

$\sim 1 \times 10^8 \text{M}^{-1}\text{s}^{-1}$ at $25^\circ$C, but may be appreciably lower in the solid state. In comparison $k_2$ for oxygen competition for the alkyl radical is $2 \times 10^9 \text{M}^{-1}\text{s}^{-1}$. Thus for air-saturated PPH ($[\text{O}_2] \sim 8 \times 10^{-4}$M) reaction 7 will be $\leq 1/160$ of the rate of reaction 2. To further investigate the importance of nitroxides in UV stabilization, 4-oxo-2,2,6,6-tetramethylpiperidine-N-oxyl was added to PPH. Photo-protection against xenon irradiation was improved as compared to the parent piperidine by about 25%, but the nitroxide itself was reduced to the $1 \times 10^{-4}$M level within the first 100h and persisted at this level until brittle failure (7). In contrast the parent amine is completely destroyed in the first 100h of xenon exposure.

The formation of appreciable quantities (up to $\sim 80\%$ based on the initial additive concentration) of the grafted substituted hydroxylamine (>NOPP as from reaction 7) in photo-degrading PPH can be demonstrated by indirect methods (10, 11). For example after the rapid loss of the initial concentration of a piperidine or its nitroxide in PPH film, heating the film immersed in iso-octane for several hours at $100^\circ$C in the presence of oxygen causes the re-appearance of nitroxide in appreciable quantities as measured by e.s.r. spectroscopy (11). This nitroxide most likely results from a reaction analogous to reaction 8 (12). In addition we have observed the >N-O-C band (at 1306 cm$^{-1}$) in the infrared spectrum of irradiated, nitroxide-containing PP films by Fourier Transform IR spectroscopy (11).

The substituted hydroxylamine from reaction 7 ( $>$NOPP) may also contribute to the stabilization process by scavenging peroxyl radicals (reaction 9).  From a study of model compounds in

$$>NOPP + PPO_2 \cdot \xrightarrow{k_9} \ >NO\cdot + [PPOOPP] \qquad\qquad 9$$

the liquid phase, $k_9/k_3$ is expected to be $\sim 100$ ($\underline{8}$).  If the $>$NOPP concentration is assumed to be close to that of the initial piperidine or nitroxide level added to the film, the relative rates of reaction 9 to reaction 3 will be $\sim 1:23$.  However the $>$NOPP species can in fact stabilize PPH against photo-oxidation.  This was demonstrated by quantitatively generating $>$NOPP from the $\gamma$-irradiation of 4-oxo-2,2,6,6-tetramethylpiperidine in PPH film in the absence of $O_2$.  Data from the subsequent Xe arc irradiation of this film and control samples are shown in Figure 1.

For the series of hindered amine derivatives based on 4-oxo-2,2,6,6-tetramethylpiperidine, our data indicate that UV stabilization effectiveness increases in the series $>$NH< $>$NO·< $>$NOPP for equimolar concentrations and identical irradiation conditions (Figure 1).  However this order is probably more a reflection on the relative volatility of the species than any chemical influences.  Loosely sandwiching $>$NH or $>$NO· containing films between quartz windows to reduce evaporative loss during Weather-Ometer irradiation increases the effectiveness of each additive by over a factor of 2, presumeably largely due to the increased retention of the piperidyl species as compared to free (uncovered) films.  Even the effectiveness of the $>$NOPP species is enhanced $\sim 50\%$ by sandwiching.  This is consistent with the inhibition cycle of reactions 7 and 9 involving the generation of the (mobile and volatile) nitroxide.

In recent publications, Gugumus ($\underline{6}$) and Felder et al ($\underline{13},\underline{14}$) have proposed a modification of the Boozer-Hammond mechanism ($\underline{15}$) to explain their observed kinetics in the inhibition of ketone photo-oxidations by hindered amine species (reaction 10).  Peroxyl radicals were suggested to form charge transfer complexes with the

$$>N\text{-}X + RO_2 \cdot \ \rightleftharpoons \ [>N\text{-}X \ \text{---} \ RO_2\cdot]$$

(X = -H, O· or -R')                                          $\Big\Updownarrow$ $RO_2\cdot$     10

$>N\text{-}X + O_2 +$ Peroxyl $\longleftarrow$    $[RO_2\cdot \ \text{--}>N\text{-}X \ \text{---} \ RO_2\cdot]$
   termination
   products

piperidinyl species and enhance the rate of peroxyl-peroxyl termination.  Although this mechanism can explain inhibition without (for example) nitroxide consumption, the oxidation kinetics of their model appear to be complex and incompletely characterized.  In addition the Boozer-Hammond mechanism has been shown to be invalid for aromatic amines ($\underline{16}$), so that reaction 10 obviously requires further substantiation.

## Hydroxylamine Effects

The substituted hydroxylamine ($>$NOPP from reaction 7) can take part in various dark reactions, even at ambient temperature. From a study of the low molecular weight model I in the liquid phase, two decomposition pathways are possible (reaction 8) (12). The products from the disproportionation reaction 8a were only observed in the absence of a radical trap such as $O_2$. In a given solvent $k_{8_b}/k_{8_a} \simeq 40$ (solvent air saturated and degassed respectively). Both $k_{8_a}$ and $k_{8_b}$ were found to increase by an order of magnitude on going from a non-polar solvent (iso-octane) to a polar solvent (methanol or tert.-butyl hydro peroxide, BuOOH).

Unsubstituted hydroxylamines ($>$NOH) which may be formed in reactions 6 and 8 are powerful antioxidants, both by peroxyl radical scavenging and by hydroperoxide decompositions. Chakraborty and Scott have reported the detection of $>$N-OH groups in the photo-oxidation of methylcyclohexane containing a bis-piperidine or a nitroxide, based on their observation of an IR absorption at $\sim$2765 cm$^{-1}$(13). However we have found the $>$N-OH absorption to occur at 3460 cm$^{-1}$ (11) in 1-hydroxyl-2,2,6,6-tetramethylpiperidines which throws some doubt on their IR detection of $>$NOH.

In fact the extremely rapid reaction of $>$NOH with hydroperoxides combined with the ready oxidation of hydroxylamines to nitroxides during storage even in the solid state makes unlikely the detection of $>$NOH from hindered amines in photo-oxidizing polymer.

## Effects of the Solid State on Photostabilization

Most kinetic treatments of the photo-oxidation of solid polymers and their stabilization are based on the tacit assumption that the system behaves in the same way as a fluid liquid. Inherent in this approach is the assumption of a completely random distribution of all species such as free radicals, additives and oxidation products. In all cases this assumption may be erroneous and has important consequences which can explain inhibition by the relatively slow radical scavenging processes (reactions 7 and 9) discussed in the previous section.

### Stabilizer- PPOOH Association

The photo-oxidation of a solid branched alkane can be expected to proceed in localized domains, new oxidation chains being generated from the photo-cleavage of -OOH products, and chain propagation (reactions 2 and 3) being concentrated close to each initial site in a given domain to produce a zone of high -OOH concentration. Thus the distribution of an additive in and around these domains is of special importance.

Nitroxides have been shown to associate quite strongly with model hydroperoxides in the liquid phase (17). We have now found evidence for $>$NO·/hydroperoxide group association in solid PPH films based both on the e.s.r. spectra of $>$NO· species and

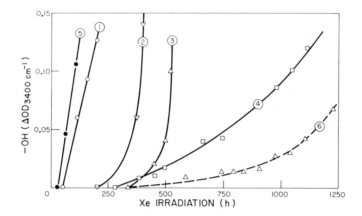

*Figure 1.    Comparative UV stabilization of PPH by piperidine species.*

*Atlas xenon arc Weather-Ometer irradiation of 25-μm film containing 4-oxo-2,2,6,6-tetra-methyl piperidine species. (1) Unstabilized PPH; (2) PPH + $>$NH (5 × 10⁻³M); (3) PPH + $>$NO·(6 × 10⁻³M); (4) PPH + PP-O-N$<$ (5 × 10⁻³M, prepared by γ-irradiation of $>$NO·/PPH under N₂); (5) PPH (pre-γ-irradiated under N₂ as for (4); (6) As for (3) but film loosely sandwiched between quartz windows.*

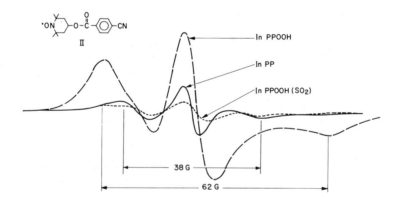

*Figure 2.    The ESR spectra of nitroxide II in photooxidized PPH.*

*Spectra recorded at 20°C with constant amplification for similar weights of PPH films. PPOOH concentration 0.10M before SO₂ treatment. In PPH, [II] = 1.5 × 10⁻⁴M; in PPOOH, [II] = 1.4 × 10⁻³M.*

on nitroxide solubility in PPH films (18). The e.s.r. spectrum of
nitroxide II (Figure 2, insert) in various PPH films has been
found to be markedly dependent on the oxidation state of the film
(Figure 2). When II is diffused into a pre-oxidized film the e.s.
r. of the dry solid sample shows both that the nitroxide level is
appreciably higher than in unoxidized film and that the nitroxide
has a very restricted motion in the oxidized polymer. The ~62G
extrema separation (Figure 2) is consistent with $>$N-O--HOOPP
association, restricting rotation about the N-O axis. This e.s.r.
spectral feature has frequently been demonstrated for nitroxide
probes in block co-polymers (two phase systems) together with a
narrower extrema separation (approaching 35G) indicative of freely
tumbling nitroxide molecules in the soft domains of the block copolymer (19). Such a narrower extrema separation (~38G) dominates
the e.s.r. spectrum of II in un-oxidized PPH (Figure 2). In
addition chemical destruction of the -OOH sites by film treatment
with $SF_4$ or $SO_2$ causes the e.s.r. spectrum of II to revert to
that found in the unoxidized polymer. The dependence of the
solubility of II on the level of oxidation of PPH is clearly shown
in Table I.

TABLE I

Effects of PPH Film Photo-oxidation

on the Solubility of Nitroxide II

| Xe Irradiation Time[a] (h) | Hydroperoxide Concentration[b] $Mx10^2$ | Nitroxide Concentration[c] $Mx10^4$ |
|---|---|---|
| 0 | <0.5 | 1.5 |
| 40 | 3.2 | 6.6 |
| 57 | 14 | 14 |
| 70 | 30 | 26 |
| 57 (+$SF_4$ or $SO_2$) | <0.5 | 0.5 |

a) Atlas 6500W Weather-Ometer.
b) Hydroperoxide concentration estimated from the 3400 $cm^{-1}$ IR
   absorption in 25 μm films.
c) All films immersed for 8h in an iso-octane solution of II
   ($3.3x10^{-2}M$) at $20°C$, then vacuum dried. Concentrations
   measured by e.s.r. spectroscopy.

In the solid state, small $>$NO· species can diffuse through
the amorphous zones, and clearly become associated with the
oxidized domains in these amorphous zones. Thus the local $>$NO·
concentration adjacent to an -OOH site (before photo-cleavage) is
anticipated to be much greater than expected from the overall $>$NO·
concentration, and so increase the effectiveness of the scavenging
process (reaction 7) as compared to the propagation step (reaction
2).

Association phenomena involving aliphatic amines are also possible. For example the association of simple amines with hydroperoxides to form stable adducts has been known for many years (20,9). Recently various 4-substituted-2,2,6,6-tetramethylpiperidines have been reported to associate strongly with tert.-butyl hydroperoxide in the liquid phase, and to form isolatable solid adducts (21). This process can be expected to occur also in PPH samples, with piperidines diffusing to and concentrating within oxidized domains containing -OOH groups. We have in fact found data analogous to that shown in Table I for various piperidines (12). The additive molecules will then be ideally placed to scavenge radical species resulting from PPOOH photocleavage (cf. reference 22) and to cause -OOH decomposition (reaction 6). In addition, we have recently found that the rate of piperidine conversion to nitroxide in preoxidized PPH is appreciably higher in solid, solvent-free polymer than in the presence of a solvent (23). This process appears to be fast enough to cause a significant loss of -OOH groups by the dark reaction, and is presumably enhanced by the solid state favoring even stronger $>$NH---HOO- association than found in the liquid phase.

### Non-uniform Distribution of Radicals

Radical distributions in the solid state can be extremely non-uniform, as compared to liquids where diffusion quickly randomizes radical populations. After the hydroperoxide photolysis (reaction 1) and attack on the PPH, the $PPO_2^{\bullet}$ pair which results from each successful initiation will be in extremely close proximity, and in the solid polymer will only slowly separate by a combination of segmental diffusion, oxidative propagation (reactions 2 + 3), etc. This phenomenon has been shown both theoretically and experimentally to result in $\sim$95% of the radical pairs undergoing rapid self-termination after only a few propagative cycles (secondary cage combination). The remaining 5% of pairs escape secondary cage combination, propagate for many thousands of steps before random termination occurs, and cause the bulk of the oxidation (24, 25). This long kinetic chain length (possibly $>10^4$) of only a small percentage of peroxyl radicals resulting from PPOOH photocleavage means that a solid state photooxidation will be much easier to retard by free radical scavenging than a liquid phase process for equivalent rates of initiation and re-activities because of the high kinetic chain length of the freely propagating radicals in the solid state. Thus low $k_7/k_2$ and $k_9/k_3$ ratios can still be expected to lead to effective inhibition of the polymer photo-oxidation.

### $>$NOPP Decomposition

From a study of the thermal stability of $>$NOPP (formed by $\gamma$-irradiation of $O_2$-free PPH + $>$NO·) we have observed the regeneration of $>$NO· for films stored in the dark, presumably by reaction 8. However only when tert.-butyl hydroperoxide was diffused into the film did we observe a relatively rapid genera-

tion of $\geq$NO· ($k_8 \approx 6 \times 10^{-7} M^{-1}s^{-1}$ at 25°C, Figure 3) ($\underline{11}$). Nevertheless reaction 8a to give $\geq$NOH may be greatly favored in the
rigid polymer matrix as compared to the liquid phase, where radical separation to allow reaction 8b can occur. This was indicated by the slow generation of $\geq$NO· when $\geq$NOPP was stored in
air, although no $\geq$NO· was formed during prolonged storage in N₂
(Fig. 3). Subsequent addition of BuOOH to these latter samples
caused a burst of $\geq$NO· generation (in each case the $\geq$NO· level
jumped to ~0.3 x $10^{-3}$M). The higher $\geq$NO· evolution for the N₂
stored sample is consistent with the sequence analogous to
reaction 8a:-

$$PP-O-N\!\!<\xrightarrow[25°C]{\text{Air or } N_2} \;>\!\!C\!\!=\!\!C\!\!<\; + \;>\!\!NOH \xrightarrow[\text{in air}]{BuOOH} \;>\!\!NO·$$
$$\downarrow$$
$$\text{some } >\!\!NO·$$

An unsubstituted hydroxylamine is a powerful hydroperoxide de-
composer and peroxyl radical scavenger, and could play an
important role in photo-stabilization even if present at only
a low concentration after dark intervals.

## Photostabilization by Complex Hindered Amines

The commercial hindered amines are based on several piperidine moities coupled in one molecule, for example via a bi-
functional ester through the 4-position ($\underline{4}$), or as substituents
(again via the 4-position) on triazine structures ($\underline{5}$). One
simple result of the high molecular weight of these additives
is that the parent stabilizer is much more resistant to loss by
migration/volatilization, especially during heat treatments of
the PPH articles. In addition their multifunctional nature increases the possibility of the additive becoming rapidly grafted
to the polymer backbone.

In some simple but elegant experiments, Hodgeman has demonstrated that, for the additive bis (2,2,6,6-tetramethyl-4-
piperidinyl) sebacate (III), nitroxide groups are generated during photo-protection of PPH and that some of this nitroxide cannot be solvent extracted ($\underline{26}$). This indicates that the grafted
species IV may well play a role in photo-stabilization. Although
we have also observed this species during the photo-oxidation of

$$PP-ON\!\!<\!\!\bigcirc\!\!>\!\!-O-\overset{O}{\underset{||}{C}}-(CH_2)_8-\overset{O}{\underset{||}{C}}-O-\!\!<\!\!\bigcirc\!\!>\!\!NO·$$
$$IV$$

PPH in the presence of III (or its bis-nitroxide) we have found
that the doubly grafted species V is generated in much larger
quantities ($\underline{23}$),

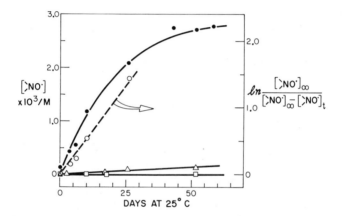

*Figure 3.    Decomposition of PP-O-N< in PPH Film During Dark Storage.*

*Initial [PP-O-N<] = 3 × 10⁻³M, prepared by γ-irradiating 4-oxo-2,2,6,6-tetramethyl-piperidine-N-oxyl in O₂-free PPH film. [>NO·] by ESR spectroscopy. (□) PP-O-N < stored under N₂; (△) PP-O-N< stored under air; (●, ○) PP-O-N < immersed in tert-butyl hydroperoxide.*

Hodgeman has suggested that the predominant extractable pro-

PP-O-N⟨⟩-O-C-(CH$_2$)$_8$-C-O-⟨⟩N-O-PP

V

duct from photo-oxidized PPH containing III is the mononitroxide
VI rather than the bis-nitroxide from III (6). This conclusion
was based on the differences between the e.s.r. spectra of the

·O-N⟨⟩-O-C-(CH$_2$)$_8$-C-O-⟨⟩N-H

VI

pure bis-nitroxide and the spectra of extracts. However the e.s.r.
spectrum of the bis-nitroxide is exceptionally sensitive to slight
changes in solvent polarity and viscosity (27, 23). We have concluded that the e.s.r. of extracts from photo-oxidized PPH film
containing III is consistent with the bis-nitroxide being the
dominant extractable product (23).

Although the possibilities of $>$NH---HOO- and $>$NO·---HOO-
association appear to be highly desirable to enhance the effectiveness of hindered amines as photo-stabilizers, it is difficult
to see how these processes can contribute to stabilization by the
macro-hindered amine additives (5). These compounds appear to be
too large to diffuse through the PPH amorphous domains at ambient
temperatures to oxidation sites where association could occur.
However association between the hindered amine stabilizers and
any oxidation products may occur in the melt when mobility is high
and persist through into the fabricated articles. Alternatively
a mechanism largely involving radical site (either -C· or PPO$_2$·)
migration to the immobile additive appears more likely. Radical
migration may occur by a combination of segmental motion, oxidative propagation (reactions 2 and 3) or hydrogen atom transfer
(28). Although this process will leave a train of PPOOH groups,
termination with the hindered amine or its products will prevent
generation of many thousands of PPOOH groups before peroxyl-
peroxyl termination occurs. If this latter mechanism is correct
the highest effectiveness of the macro hindered amines will be
achieved when the additive is uniformly dispersed throughout the
oxidizing, non-crystalline parts of the polymer.

The formation of a nitroxide-peroxyl complex (reaction 10)
(6, 13, 14) would appear to be an attractive mechanism for macro
stabilizers provided that peroxyl propagation is prevented. However further evidence of this association is required.

Conclusions

Substituted tetramethylpiperidines and their oxidation products appear to act as UV stabilizers for polypropylene by weakly
scavenging PP·, PPO$_2$· and possibly other radicals involved in the
photo-oxidation. Their weak scavenging ability is off-set by the

tendency of the parent piperidine and its nitroxide to associate
with oxidation products, the very sites where free-radical generation will occur, as well as the high kinetic chain length anticipated for freely propagating radicals in PPH photo-oxidations.
Hydroxylamine liberation in a dark reaction may aid peroxyl
scavenging and -OOH decomposition.  The multifunctional commercial
piperidines mainly offer advantages over the simple mono-
piperidines because of their resistance to migration to the surface and evaporative loss.

## Literature Cited

1. Carlsson, D.J. and Wiles, D.M., J. Macromol. Sci., Rev.
   Macromol. Chem., 1976, C14, 65.
2. Niki, E., Decker, C., and Mayo, F.R., J. Polym. Sci., Polym.
   Chem. Ed. 1973, 11, 2813.
3. Carlsson, D.J., and Wiles, D.M., J. Macromol. Sci., Rev.
   Macromol. Chem. 1976, C14, 155.
4. Usilton, J.J., and Patel, A.R., Adv. Chem. Ser. 1978, 169, 116
5. Tozzi, A., Cantatore, G., and Masina, F., Text. Res. J., 1978
   48, 433.
6. Gugumus, F., Chap. 8 in "Developments in Polymer Stabilization"
   vol. 1, ed. G.Scott, Applied Science Publishers, London 1979
7. Carlsson, D.J., Grattan, D.W., Suprunchuk, T., and Wiles, D.M.
   J. Appl. Polym. Sci., 1978, 22, 2217.
8. Grattan, D.W., Carlsson, D.J., and Wiles D.M., J. Polym. Deg.
   and Stability, 1979, 1, 69.
9. Hiatt, R., Chapter 1 in "Organic Peroxides", Vol.II, ed.
   Swern, D., Wiley N.Y. (1971).
10. Shlyapintokh, V.Y., Ivanov, V.B., Khvostach, O.M., Shapiro,
    A.B., and Rozantsev, E.G., Dokladi Akad. Nauk. SSSR, 1975, 225
    1132.
11. Chan, K.H., Carlsson, D.J., and Wiles, D.M.. unpublished
    results.
12. Grattan, D.W., Carlsson, D.J., Howard, J.A., and Wiles, D.M.,
    Can. J. Chem. 1979, 57, 2834.
13. Felder, B., Schumacher, R., and Sitek, F., Helv. Chim. Acta,
    1980, 63, 132.
14. Felder, B., Schumacher, R., and Sitek, F., Amer. Chem. Soc.
    Org. Coatings Plast. Prep.(Houston), 1980, 42, 561.
15. Boozer, C.E., Hammond, G.S., Hamilton, C.E., and Sen, J.N.,
    J. Amer. Chem. Soc., 1955, 77, 3233.
16. Adamic, K., Bowman, D.F., and Ingold, K.U., J. Amer. Oil Chem.
    Soc. 1970, 47, 109.
17. Grattan, D.W., Reddoch, A.H., Carlsson, D.J., and Wiles, D.M.
    J. Polym. Sci. Polym. Letters Ed., 1978, 16, 143.
18. Chan, K.H., Carlsson, D.J., and Wiles, D.M., J. Polym. Chem.
    Polym. Lett. Ed. in press.
19. Kumler, P.L. Keinath, S.E., and Boyer, R.F., Polym. Eng. Sci.
    1977, 17, 613.

20. Oswald, A.A., Hudson, B.E., Rodgers, G., and Noel, F., J. Org. Chem. 1962, 27, 2439.
21. Sedlar, J., Petruj, J., Pac, J., and Navratil, M., Polymer 1980, 21, 5.
22. Sedlar, J., Petruj. J., Pac, J., Zahradnickova, A., Euro. Polym. J. 1980, 16, 663.
23. Durmis, J., Chan, K.H., Carlsson, D.J., and Wiles, D.M., Unpublished results.
24. Carlsson, D.J., Chan, K.H., Garton, A., and Wiles, D.M., Pure Appl. Chem. 1980, 52, 389.
25. Garton, A., Carlsson, D.J., and Wiles, D.M., Die Makromol. Chem., 1980, 181, 1841.
26. Hodgeman, D.K., J. Polym. Sci. Polym. Chem. Ed. 1980, 18, 533.
27. Buchachenko, A.L., Golubev, V.A., Medzhidov, A.A., Rozantsev, E.G., Teor. Eksp. Khim 1965, 1, 249.
28. Denisov, E.T., Uspekhi Khimii, 1978, 47, 1090.

RECEIVED October 14, 1980.

# Hindered Amine Light Stabilizers

## A Mechanistic Study

B. FELDER, R. SCHUMACHER, and F. SITEK

Plastics and Additives Division, CIBA–GEIGY Limited, 4002 Basel, Switzerland

## Synopsis

Mechanisms for the action of hindered amine light stabilizers are reviewed briefly. On the basis of these
considerations, two aspects of the mode of action of
these substances were examined more closely.

First the interaction of selected tetramethylpiperidine
(TMP) derivatives with radicals arising from Norrish-
type I cleavage of diisopropyl ketone under oxygen was
studied. These species are most probably the isopropyl
peroxy and isobutyryl peroxy radicals immediately formed
after α-splitting of diisopropyl ketone and subsequent
addition of $O_2$ to the initially generated radicals.
Product analysis and kinetic studies showed that the investigated TMP derivatives exercise a marked controlling
influence over the nature of the products formed in the
photooxidative process. The results obtained point to
an interaction between TMP derivatives and especially
the isobutyryl peroxy radical.

Secondly, the interaction of hindered amines with hydroperoxides was examined. At room temperature, using different monofunctional model hydroperoxides, a direct
hydroperoxide decomposition by TMP derivatives was not
seen. On the other hand, a marked inhibitory effect of
certain hindered amines on the formation of hydroperoxides in the induced photooxidation of hydrocarbons was
observed. Additional spectroscopic and analytical evidence is given for complex formation between TMP derivatives and tert.-butyl hydroperoxide. From these results,
a possible mechanism for the reaction between hindered
amines and the oxidizing species was proposed.

## Introduction

Light and oxygen cause more or less pronounced degradation in almost all unstabilized plastics. It is therefore common practice to add stabilizers to a polymeric
material that multiply its resistance to photooxidation
by many times[2-5]. Most of the light stabilizers so far
used can be assigned by their mode of action to one of
four groups: light screeners, UV absorbers, quenchers
and a group comprising hydroperoxide decomposers and
radical scavengers. Their various modes of action have
been discussed in detail in a number of publications
[6-12].

In the present study possible modes of action of a relatively new class of light stabilizers, the Hindered
Amine Light Stabilizers (HALS, tetramethylpiperidine
derivatives, TMP) will be discussed. Certain members
of this group have a light stabilizing effect superior
in many fields of application to that of previous
additives. They are effective in both thick layers and
fibers[13-19].

The light stabilizing action of these substances has
been studied mainly in polyolefins. A typical finding
in one of the earlier studies made in our application
laboratories by Leu / Gugumus is illustrated in Figure 1
which shows the formation of carbonyl groups as a function of the irradiation time in the Xenotest 150 of PP
sheets stabilized in various ways[13,14]. Of the light
stabilizers tested, a HALS derivative (Bis-(2,2,6,6-
tetramethyl-4-piperidinyl)sebacate, HALS 1) conferred
longest life on the samples. The very marked activity
of these compounds is not confined, however, to polyolefins but extends to other plastics such as styrene
polymers, polyurethanes and polydienes[20-23].

An aspect of particular interest is the performance of
hindered amines in protective coatings such as lacquers. Weathering tests in our application laboratories
have shown that a tetramethylpiperidine derivative
(HALS 1) has in fact a very good protective effect.
Fig. 2 demonstrates the light stability of a high-solid
lacquer based on two-component polyurethane. The loss
of gloss was measured as a function of the irradiation
time in a QUV instrument (Q-Panel Company, Cleveland
Ohio). The figure also shows that a particularly good

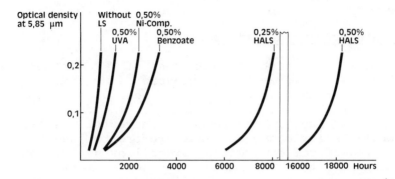

*Figure 1.   Formation of carbonyl groups in 0.1 mm PP films as a function of irradiation time in Xenotest 150 (13). (UVA) 2-(2'-hydroxy-3',5'-di-tert-butylphenyl)-5-chloro-benzotriazole; (Ni-Comp) Nickel[2,2'-thiobis-(4-tert-octylphenolate)]-n-butyl-amine; (Benzoate) 2,4-di-tert-butylphenyl-3,5-di-tert-butyl-4-hydroxy-benzoate; (HALS) Bis-(2,2,6,6-tetramethyl-piperidinyl-4)sebacate.*

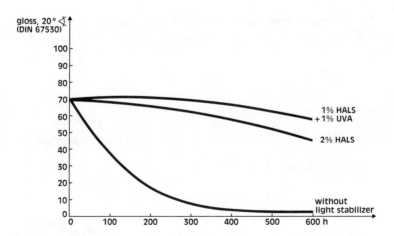

*Figure 2.   Light stability of a high-solids, one-coat silver metallic paint based on a two component polyurethane. Test criterion: gloss retention; QUV irradiation. (HALS) Bis-(2,2,6,6-tetramethyl-piperidinyl-4)sebacate; (UVA) 2-(2'-hydroxy-3',5'-di-tert-amylphenyl) benzotriazole.*

protective effect is reached with the combination UV
absorber + hindered amine.


## Mechanistic Considerations

Considering the high effectiveness of TMP derivatives
as light stabilizers it is not surprising that the mode
of action of these substances has been widely investigated over the last few years[24-36]. Hindered amines do
not absorb light in the range of terrestrial solar radiation. An obvious step was therefore to study their
quenching effect on excited carbonyl or singlet oxygen.
Our experiments - like those of other workers - showed,
however, that hindered piperidines and their $\geq$N-methyl
derivatives do not quench excited ketones, whereas
nitroxides have this ability[25-27,37,38]. As regards
singlet oxygen quenching, it has already been shown
that only $\geq$N-methyl and nitroxide derivatives are active in this respect[25,26]. Certain secondary hindered
amines - in practice excellent light stabilizers - are
practically inactive. It follows that singlet oxygen
quenching plays at most a secondary role.

Three main possible modes of action of hindered amines
as light stabilizers have recently come under discussion:

1. The postulated interaction with hydroperoxides[30] was
   based on the observation that HALS 1 appears to be
   less active in the presence of substances known to
   decompose hydroperoxide, like nickel dibutyldithio-
   carbamate.

$$\text{piperidine structure} + ROOH \xrightarrow{\text{dark}} \text{products} \qquad 1$$

2. A further effect of sterically hindered amines that
   could contribute to polymer stabilization has been
   described by Allan / McKellar[31,39-41]. They reported
   a conversion of $\alpha,\beta$ to $\beta,\gamma$ unsaturated ketones in
   polyolefins and assumed that this played a role in
   the photooxidation of the polymers. The conversion

is suppressed in the presence of sterically hindered piperidines:

$$-\overset{|}{C}H-CH{=}CH-\overset{\overset{\displaystyle O}{\|}}{C}- \xrightarrow[\substack{{>}NH,\ {>}NO\cdot}]{h\nu/O_2} \quad\diagup\!\!\!\!\times\diagdown\qquad\qquad 2$$

$$-\overset{|}{C}{=}CH-CH_2-\overset{\overset{\displaystyle O}{\|}}{C}-$$

3. Certain TMP derivatives act as scavengers of the radicals arising in polymer degradation which was considered fairly early on to be a further possible mode of action[28,32,34,35]. In addition to the known reaction of nitroxides with alkyl radicals[24], their possible formation from the amines via >N· as intermediate step has been suggested[35]. Capture of alkyl radicals by >NO· gives rise to ethers >NOR. The subsequent reaction of these with peroxy radicals and regeneration of nitroxide is seen as a possible step in polymer stabilization[28] (reaction (3)):

$$>NH \xrightarrow{\text{radicals}} >N\cdot \xrightarrow{\text{"}O_2\text{"}} >NO\cdot \xrightarrow{R\cdot} >NOR$$

$$\downarrow R'OO\cdot \qquad 3$$

$$>NO\cdot + ROOR'$$

However, a comparison made by Carlsson of the experimentally obtained protection effects and those calculated on the basis of reaction (3) showed that process (3) cannot be alone responsible for the observed stabilizing effect[34].

As previous investigations have shown[26,37] a study of the "classical" protective mechanisms does not contribute much towards explaining the mode of action of HALS derivatives. An obvious next step therefore seemed to be a closer investigation of the action of these substances on radicals and hydroperoxides.

Opinions differ as to the nature of the species primarily responsible for photooxidation of polyolefins[42-47]. Many workers put the main responsibility for photodegradation on hydroperoxides and their breakdown products. Others, like Guillet for example, consider the ketone groups to be much more important, at least where long-term behavior of polymers is concerned. In Guillet's view, these are responsible for cleavage of the polymer chains and thus also for the change in the physical properties of polymers[46,47].

Ketone photolysis in an inert atmosphere has been widely studied[48-50]. Apart from polymer photooxidation studies, however, little work has been done on their degradative irradiation in an oxidizing medium[51-53]. For this reason we have concentrated on the study of ketone photolysis in the presence of oxygen and the interaction of the oxygen-centered radicals arising in this reaction with certain tetramethylpiperidine derivatives.

In previous experiments[1] dibenzyl ketone was used as model substance. This compound undergoes exclusively Norrish type I cleavage with a high quantum yield[54,55]. Its photolysis in the presence of oxygen gives rise to a heterogeneous product mixture - not a surprising result where radical reactions under oxygen are concerned (reaction (4)):

$$PhCH_2\overset{\overset{\text{O}}{\|}}{C}CH_2Ph \xrightarrow{h\nu/O_2} PhCHO + PhCH_2\overset{\overset{\text{O}}{\|}}{C}OH + Ph\overset{\overset{\text{O}}{\|}}{C}OH \qquad 4$$

DBK                                                    $+ PhCH_2OH$

$$\text{yield} \sim 50\%$$

$$PhCH_2\overset{\overset{\text{O}}{\|}}{C}CH_2Ph \xrightarrow[\text{TMP derivatives}]{h\nu/O_2} PhCHO + PhCH_2\overset{\overset{\text{O}}{\|}}{C}OH \qquad 5$$

In the presence of TMP derivatives, on the other hand, a controlling influence on the course of the reaction appears: only two products are formed, and these in amounts bearing a stoichiometric relationship to the amount of DBK decomposed (reaction (5)). This surprising effect of HALS derivatives has not previously been reported. Therefore the question arises: is this

a special case or does it apply to ketones in general?
In particular one has to ask whether it applies to pu-
rely aliphatic ketones, whose structure is more similar
to that of the carbonyl groups assumed to be present in
polyolefins. For the present study diisopropyl ketone
(DiPK), which undergoes also only Norrish type I clea-
vage with a quantum yield of $\Phi_{CO,50}o = 0.93\underline{56}$ was chosen.

Results and Discussion

Diisopropyl ketone photolysis under exclusion of oxygen

In the presence of the nitroxide I, DiPK photolysis
yields - in amounts equivalent to the loss of ketone -
the two combination products of nitroxyl with the isopropyl and isobutyryl radical respectively (Fig. 3,
reaction (6)):

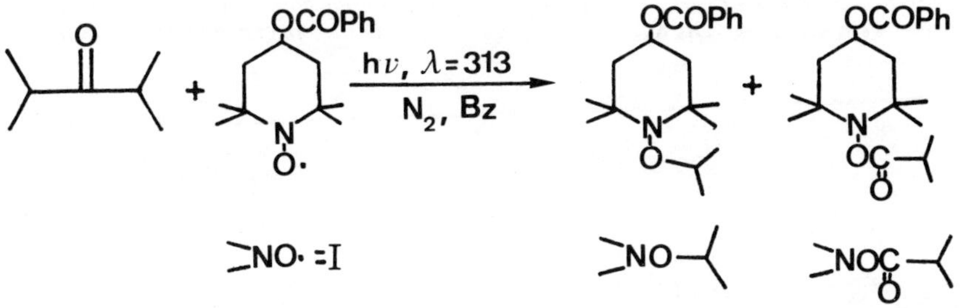

To render the time scale independent of the absorption
and physical quenching of the reaction partners, the
concentrations during the reaction were plotted against
the loss of ketone, i.e. against

$$1 - \frac{[DiPK]_t}{[DiPK]_o} .$$

As can be seen from Figure 3, the ratio of the isopropyl ether to isobutyrate is about 1:1. It is clear that
after α-cleavage of the ketone the two radicals primarily formed are captured directly by nitroxide. This
takes place without decarbonylation of the acyl radical
(reaction (7)):

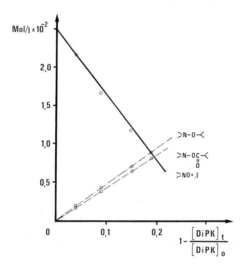

*Figure 3.   Photolysis ($\lambda = 313$ nm) of DiPK in the presence of $N_2$ and nitroxide I; solvent: benzene; $1 - [DiPK]_t/[DiPK]_o$: decrease in DiPK concentration; $[DiPK]_o = 5 \times 10^{-2}$ mol/L*

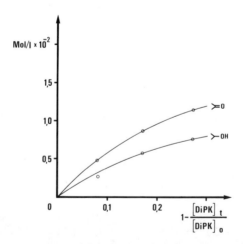

*Figure 4.   Photolysis ($\lambda = 313$ nm) of DiPK in the presence of $O_2$; solvent: benzene; $[DiPK]_o = 5 \times 10^{-2}$ mol/L*

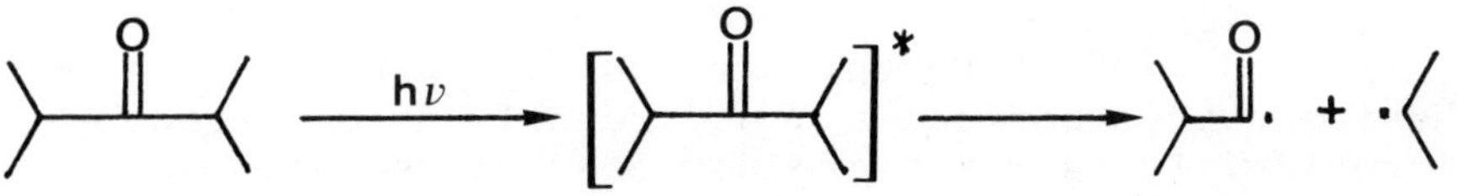

$$\text{)}\!\overset{O}{\underset{}{\|}}\!\text{(} \xrightarrow{\ h\nu\ } [\text{)}\!\overset{O}{\underset{}{\|}}\!\text{(}]^* \longrightarrow \text{)}\!\overset{O}{\underset{}{\|}}\!\cdot \ +\ \cdot\text{(}$$

a       b

7

## Diisopropyl ketone irradiation in the presence of oxygen

From the foregoing degradation scheme for DiPK in the presence of nitroxide, as well as from what we know about DBK photooxidation, it would be expected that the two radicals a and b primarily formed would be captured by oxygen and then give rise to the isobutyryl and iso-propyl peroxy radicals c and d. Theoretically, these could interact to form isobutyric acid and acetone (reaction (8)):

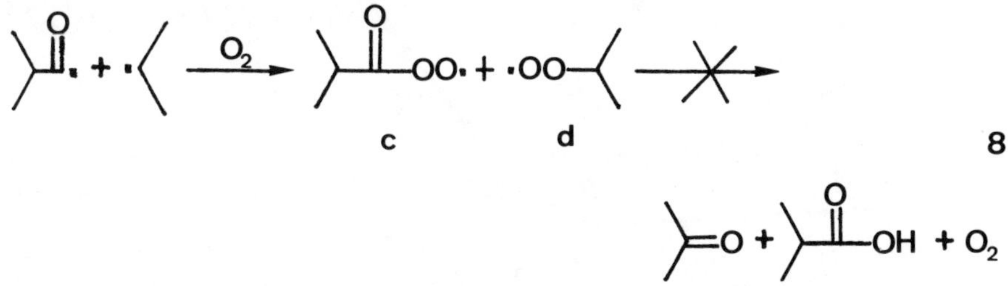

$$\text{)}\!\overset{O}{\underset{}{\|}}\!\cdot\ +\ \cdot\text{(} \xrightarrow{\ O_2\ } \text{)}\!\overset{O}{\underset{}{\|}}\!\text{OO}\!\cdot\ +\ \cdot\text{OO}\!\text{(} \xrightarrow{\ \ \ } \!\!\!\!\times$$

c      d               8

$$\text{)=O}\ +\ \text{)}\!\overset{O}{\underset{}{\|}}\!\text{OH}\ +\ O_2$$

Actually one observes (Fig. 4) the formation of acetone and isopropanol, with only traces of isobutyric acid (reaction (9), both products being formed in amounts exceeding 80 % of the amount of DiPK decomposed as indicated by the initial rate of formation).

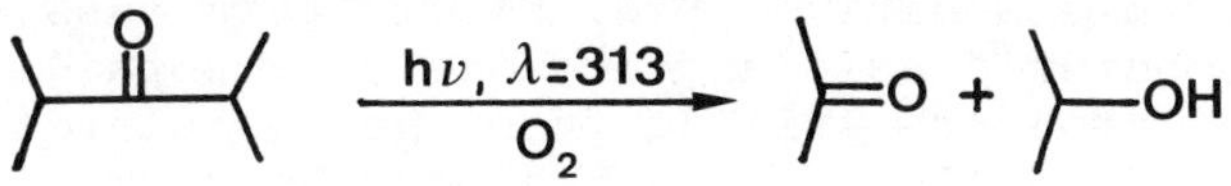

$$\text{)}\!\overset{O}{\underset{}{\|}}\!\text{(} \xrightarrow[\ O_2\ ]{\ h\nu,\ \lambda=313\ } \text{)=O}\ +\ \text{)}\!\text{OH} \qquad\qquad 9$$

A possible explanation for this is decarbonylation of the isobutyryl radical prior to its further reaction with oxygen (reaction (10)):

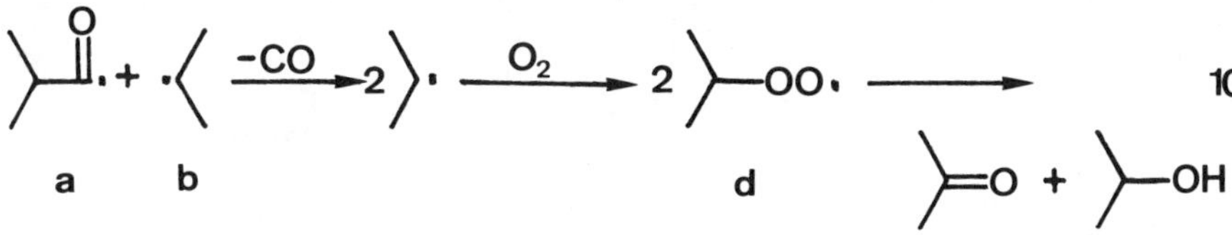

$$\text{)}\!\overset{O}{\underset{}{\|}}\!\cdot\ +\ \cdot\text{(} \xrightarrow{\ -CO\ } 2\ \text{)}\!\cdot \xrightarrow{\ O_2\ } 2\ \text{)}\!\text{OO}\!\cdot \longrightarrow \qquad 10$$

a     b                 d

$$\text{)=O}\ +\ \text{)}\!\text{OH}$$

The reactions of radicals with oxygen are diffusion-controlled[57,58]. Moreover, as has been previously shown, the isobutyryl radical a could readily be captured by a nitroxide. It is therefore not easy to see why reaction between oxygen and the species a does not also occur.

Actually, it is much more likely that two acylperoxy radicals combine with liberation of oxygen and $CO_2$[59] to form two isopropyl radicals which would then react with oxygen to yield isopropylperoxy radicals (reaction (11)):

Diisopropyl ketone photolysis in the presence of oxygen and ⊃NO · as additive
_____

In the presence of nitroxide I, diisopropyl ketone photooxidation takes a course differing considerably from that without this additive (Fig. 5). In this case high yields of isobutyric acid and acetone were obtained, presumably as products arising from the postulated peroxy radicals c and d. On the other hand, the formation of isopropanol is almost completely suppressed.

If one takes into account not only the initial slope of
the curves but also the part played by the formation
of isobutyrate it can be seen that the amount of reaction products formed is almost equivalent to the loss
of DiPK. In this case the formation of isobutyric acid
represents the most important difference compared with
irradiation without additive. It shows that in the presence of nitroxide the acyl radical may not only be
captured by oxygen but can also react further as acylperoxy radical, without losing its carbonyl group in
the process.

An interesting finding is that the product distribution
is the same when the ratio of $[\geq NO \cdot]_0$ to $[DiPK]_0$ is
only 1 : 10 (Fig. 6, in Fig. 5 $\sim$ 1 : 2), that is to
say when the nitroxide represents only a fraction of the
total amount of DiPK decomposed.

It will also be apparent from the two preceding figures
(5 and 6) that during the course of the reaction the
decrease in the amount of nitroxide present is relatively small - this despite the fact that the additive
has a decisive influence on the whole reaction.

In any discussion of a possible mechanism of the above
reaction two findings appear to be of importance:

1. The change in the product mixture in the presence of
   nitroxide I, i.e., formation of isobutyric acid instead of isopropanol, and

2. The fact that the nitroxyl radical emerges from the
   reaction practically unchanged points to a mechanism
   in which there is specific regeneration of nitroxide
   involving all the radicals present.

In analogy with our earlier experiments using dibenzyl
ketone, there are here again two possible courses of
the reaction:

1. Formation of a Charge Transfer Complex between the
   nitroxide and one of the oxygen-centered radicals
   c or d. In the next step the complex is assumed to
   react with the remaining peroxyradical to form the
   products actually found, with release of the additive (reactions (13)/(14) or (13')/(14')):

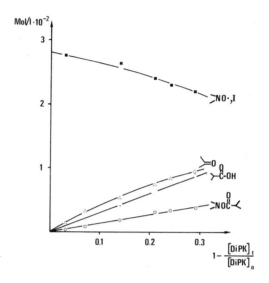

*Figure 5.    Photolysis (λ = 313 nm) of DiPK in the presence of $O_2$ and nitroxide I;
solvent: benzene; $[DiPK]_o = 5 \times 10^{-2}$ mol/L*

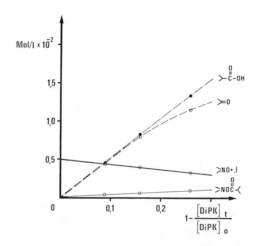

*Figure 6.    Photolysis (λ = 313 nm) of DiPK in the presence of $O_2$ and nitroxide I;
solvent: benzene; $[DiPK]_o = 5 \times 10^{-2}$ mol/L*

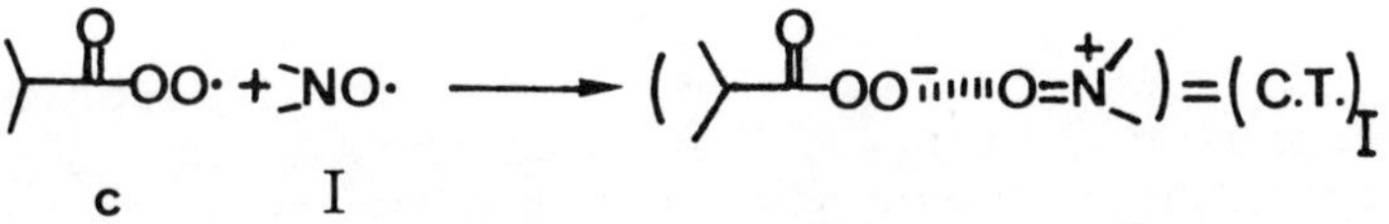

$$\text{\raisebox{0pt}{}}\text{OO}\cdot + {:}NO\cdot \longrightarrow \left( \text{\raisebox{0pt}{}}\text{OO}^{-}\cdots\cdots O{=}\overset{+}{N}{<} \right) = (C.T.)_{I} \qquad 13$$

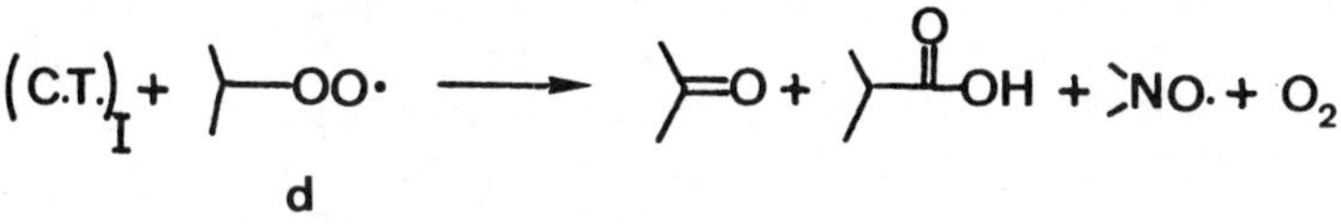

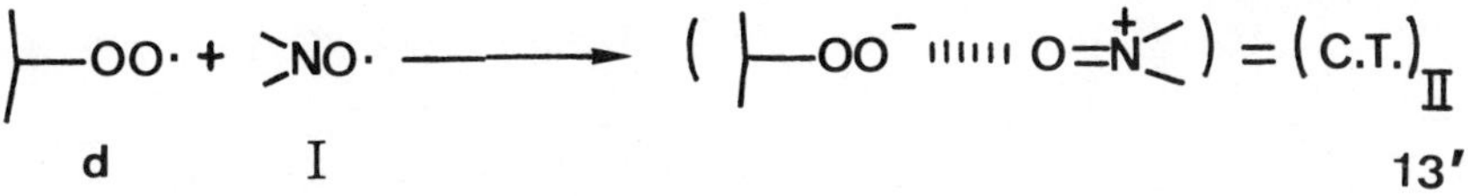

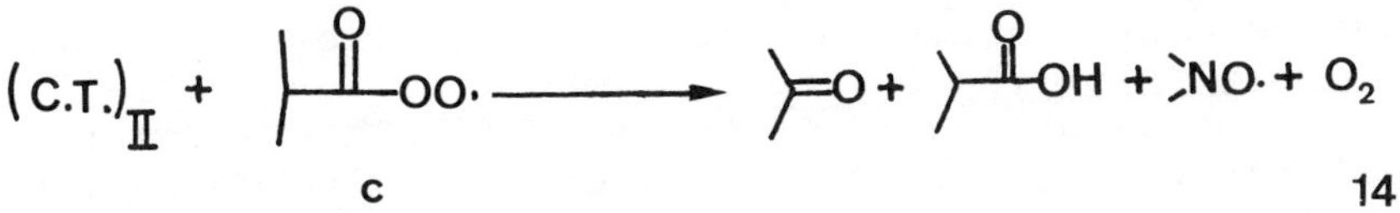

c  I

$$(C.T.)_{I} + \text{\raisebox{0pt}{}}{-}OO\cdot \longrightarrow {>}{=}O + \text{\raisebox{0pt}{}}{-}OH + {:}NO\cdot + O_2 \qquad 14$$

d

$$\text{\raisebox{0pt}{}}{-}OO\cdot + {>}NO\cdot \longrightarrow \left( \text{\raisebox{0pt}{}}{-}OO^{-}\cdots\cdots O{=}\overset{+}{N}{<} \right) = (C.T.)_{II} \qquad 13'$$

d  I

$$(C.T.)_{II} + \text{\raisebox{0pt}{}}{-}OO\cdot \longrightarrow {>}{=}O + \text{\raisebox{0pt}{}}{-}OH + {:}NO\cdot + O_2 \qquad 14'$$

c

2. A second mechanism involving as intermediate step a stable hydroxylamine ether (isopropyl I-ether) is also a possibility (reaction (15)). In a second step the ether would undergo cleavage by the acylperoxy radical with formation of isobutyric acid and acetone and liberation of the nitroxide (reaction (16)):

$$\text{\raisebox{0pt}{}}{-}OO\cdot + {>}NO\cdot \longrightarrow {>}{-}ON{<} + O_2 \qquad 15$$

d  I

$$\text{\raisebox{0pt}{}}{-}ON{<} + \text{\raisebox{0pt}{}}{-}OO\cdot \longrightarrow {>}{=}O + \text{\raisebox{0pt}{}}{-}OH + {:}NO\cdot \qquad 16$$

c

In the case of DiPK our experimental findings have not enabled us to decide between these two possibilities. However, according to the work of Rosantsev[60] and Ingold[61] reactions between peroxy radicals and nitroxides derived from tetramethylpiperidine are not very probable, so reaction (15) should not be important.

The observed formation of isobutyrate (Figs. 5 and 6) would appear to be one of the possible reasons for the slow decrease in the nitroxide concentration. The formation of isobutyrate can be seen as a reaction competing with the capture of the acyl radicals by oxygen. The absence of isopropyl ether in the reaction mixture is explained by its immediate cleavage - following its formation analogous to isobutyrate - to nitroxide by oxygen-centered radicals (mainly acyl peroxy radicals).

### DiPK photolysis in the presence of oxygen and of amine as additive

Irradiation of diisopropyl ketone under oxygen in the presence of the hindered piperidine II likewise results in formation of isobutyric acid, acetone and small amounts of isopropanol. At the same time the amine is quantitatively oxidized to the corresponding nitroxide I (Fig. 7, reaction (17)):

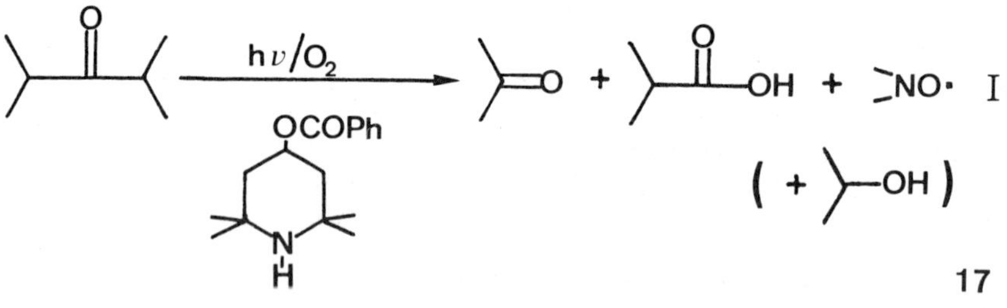

It has already been shown in DBK photooxidation[1] that of the two peroxy radicals arising under oxygen mainly one is responsible for oxidation of the amine: the acyl peroxy radical.

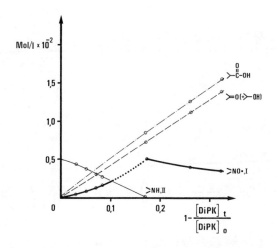

*Figure 7.    Photolysis (λ = 313 nm) of DiPK in the presence of O₂ and amine II; solvent: benzene; [DiPK]ₒ = 5 × 10⁻² mol/L*

In the case of DiPK, we were likewise able to show in
additional experiments that in all probability it is
the isobutyryl peroxy radical c and not the isopropyl
peroxy radical d that is responsible for oxidation of
the amine II to the nitroxide I. When namely the two
oxygen-centered radicals are produced independently of
one another in accordance with reactions (19) and (20)
only in the case of the acylperoxy radical the formation of the nitroxide can be observed:

$$\text{Effect of HALS on oxidation of ketones (conclusions)}$$

The experiments described in this and in the earlier
paper[1] show that the HALS derivatives studied exercise
a marked controlling influence on the product mixture
resulting from the photolysis of ketones in the presence of oxygen. A detailed study of the results leads to
the further following conclusions:

1. The formation of isobutyric acid in the presence of
   the additives studied, and the results of additional
   studies (di-tert.-butyl peroxyoxalate/isobutyroal-
   dehyde/amine), point to the intermediate formation
   of acyl peroxy radicals.

2. Kinetic analysis of the results of ketone oxidation in the presence of amine II reveals that the velocity constant of the oxidation of amines by acyl peroxy radicals must be greater (by a factor of 2 - 3) than that of the interaction of these radicals with the nitroxide[1]. In this reaction, acyl peroxy radicals are captured and destroyed by amines.

3. Also in the case of a polymer therefore, provided the acyl peroxy radicals are formed by ketone photolysis in the presence of oxygen, the oxidation of amines by these radicals would make a significantly greater contribution to stabilization than the nitroxide. The latter is in any case present in only very small amount as secondary product[1,22].

TMP derivatives and hydroperoxides
_____

As mentioned in the introduction, there are conflicting views as to the contributions made to polymer degradation by various initiating species. Among these species, in addition to ketones, hydroperoxides are some of the more important chromophores. As it is known, the photolysis of hydroperoxides yields alkoxy and hydroxy radicals. In polymers, in the presence of oxygen, these radicals lead to the secondary formation of peroxy radicals. The latter in turn are converted by hydrogen abstraction into new hydroperoxides (Scheme I):

### Scheme I

$$ROOH \xrightarrow{\quad h\nu \quad} RO\cdot \ + \ HO\cdot \qquad\qquad 21$$

$$\left.\begin{array}{c} RO\cdot \\ \\ \\ HO\cdot \end{array}\right\} + R^1H \longrightarrow \left.\begin{array}{c} ROH \\ \\ \\ H_2O \end{array}\right\} + R^1\cdot \ \xrightarrow{\ O_2\ } R^1OO\cdot \qquad\begin{array}{c} \\ \\ 22\end{array}$$

$$R^1OO\cdot \ + \ R^2H \longrightarrow R^1OOH \ + \ R^2\cdot \quad 23$$

The next step was therefore to study the interaction of TMP derivatives with peroxides and with the products arising from them during photooxidation. A direct hydroperoxide decomposing effect of the HALS derivatives studied was not observed.

Thus in mixtures with various model hydroperoxides (reaction (24)), neither amine II nor nitroxide I had any effect on the iodometrically determined peroxide content after standing for a few days at RT.

$\underset{\phantom{x}}{\overset{\phantom{x}}{+}}$—OOH

$\overset{\phantom{x}}{\underset{\phantom{x}}{\bigvee}}$—OOH + $\rangle$NH or $\rangle$NO· $\quad\dfrac{\text{RT, air}}{\text{benzene}}\quad$ →no decrease $\quad$ 24

$\phantom{xxxxxxxx}$II $\phantom{xx}$ I $\phantom{xxxxxxxxxxxxx}$ in ROOH

$\wedge\!\!\diagdown\!\!\wedge$OOH

$\phantom{xxxx}$conc. $2.5 \times 10^{-2}$ Mol/l

In the literature the possibility of oxidation of hindered piperidines by hydroperoxides has occasionally been discussed[30]. In the light of our findings at RT, however, this could at the most occur with particular hydroperoxides such as hydroperoxide sequences or α-keto, α-hydroxy and α-unsaturated hydroperoxides[62,63]. All these are quite likely to be present in oxidized polymers.

At higher temperature, however, (> 100°C) an accelerated decomposition of hydroperoxide by the hindered amine II was very clearly seen. As Table I shows, the effect is observable in solvents with differing reactivities towards radicals. It is interesting to note that in the experiments made under nitrogen the loss of amine was only a fraction of the amount of hydroperoxide decomposed. At the same time the amount of nitroxide formed was negligible ($\sim$ 1 % of the amount of amine used).

By contrast, no similar accelerating action of TMP derivatives on hydroperoxide photolysis at RT has been observed either by us or other workers[64a].

As scheme I shows, there are different possible species with which an additive of the HALS type can interact in the oxidation cycle. Thus a marked inhibitory effect of certain HALS derivatives on the formation of hydroperoxide during the photooxidation of hydrocarbons can be observed. The oxidation of 2,4-dimethylpentane, a low-molecular PP analog, was initiated by photolysis of di-tert.-butyl peroxide in the presence of oxygen (Fig. 8). The course of the oxidation was followed by monitoring the titratable active oxygen present. The method used determines the latter independently of the peroxide added. In the irradiation samples containing additive the amount of active oxygen is considerably less than in the solutions without additive. An interesting observation is that the HALS derivatives studied commence to act only when a certain concentration of hydroperoxide is built up. Similar curves are obtained when iso-octane is used as model hydrocarbon.

The experiments described above can be summarized by the following set of equations (reactions (25.1)-(25.3)):

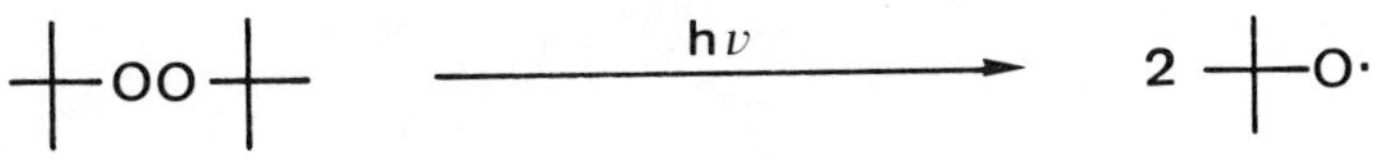

25.1

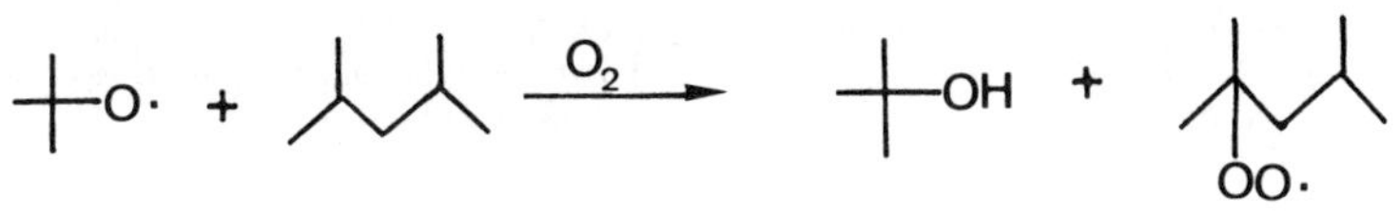

25.2

**Table I.  Thermal Decomposition of +OOH (0.05 mol/L) in the Presence of Amine II (0.05 mol/L) in a N$_2$ Atmosphere**

| Solvent | Temp. °C | Time h | Composition | % decomposed |
|---|---|---|---|---|
| benzene | 150 | 20 | +OOH | 6 |
| | 150 | 20 | +OOH + II | 39 |
| (C₆H₄)(Cl)(Cl) | 150 | 2 | +OOH | 25 |
| | 150 | 2 | +OOH + II | 90 |
| 2, 6, 10, 14-tetramethyl pentadecane | 150 | 20 | +OOH | 92 |
| | 150 | 20 | +OOH + II | 98 |
| | 120 | 20 | +OOH | 7 |
| | 120 | 20 | +OOH + II | 39 |

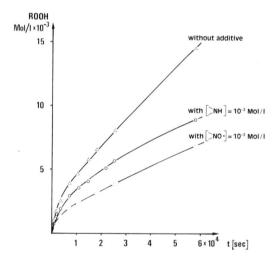

*Figure 8.  Formation of "active oxygen" as a function of irradiation time during photolysis (λ = 313 nm) of +OO+ (2 mol/L) in 2,4-dimethylpentane in the presence of O$_2$. Additives: amine II and nitroxide I*

$$25.3$$

Since the reactions occur under oxygen saturation, the principal stabilizing steps are the interactions of the HALS derivatives with the alkoxy and peroxy radicals and with the hydroperoxides.

Separate experiments in which tert.-butoxy radicals were produced thermally in benzene from di-tert.-butyl peroxyoxalate failed to reveal any direct reaction of these radicals with amine II. Even at higher temperatures ($\sim$ 150°C, dichlorobenzene, +OO+ decomposition), the +O• radicals attacked neither amine II nor nitroxide I. The earlier described experiments of ketone photooxidation showed additionally that amine II displays no specially marked reactivity towards peroxy radicals.

In sum, the results described have led us to postulate the following possible mechanism as explanation of the observed retardation of hydroperoxide formation by TMP derivatives: The HALS studied form a complex with the hydroperoxides which is much more efficiently broken down by peroxy and/or alkoxy radicals - with formation of harmless products - than hydroperoxides alone (reaction (26)). The result is a lowering of the rate of formation of hydroperoxides.

$$Y = X, \quad Y \neq X$$

26

The complex formation between hydroperoxides and HALS derivatives proposed for the preceding reaction was recently postulated by two different groups of investigators. First, Carlsson determined a complex formation constant for +OOH and a nitroxide on the basis of ESR measurements[65]. Secondly, Sedlar and his coworkers were able to isolate solid HALS-hydroperoxide complexes and characterize them by IR measurements[64a]. The accelerated thermal decomposition of hydroperoxides observed by us likewise points to complex formation. It is moreover known that amines accelerate the thermal decomposition of hydroperoxides[66]. Thus Denisov for example made use of this effect to calculate complex formation constants for tert.-butyl hydroperoxide and pyridine[67].

Complex formation constants could also be determined directly from UV spectrophotometric measurements. Addition of tert.-butyl hydroperoxide to a solution of nitroxide I in heptane at RT causes a shift of the characteristic absorption band of $\supset$NO· at 460 nm to lower wavelengths (Fig. 9). This displacement allows calculation of a complex equilibrium constant of 5 ± 1 l/Mol. Addition of amine II to the same solution causes reverse shift of the $\supset$NO· absorption band. From this one can estimate a complex formation constant for amine II and +OOH of 12 ± 5 l/Mol (23 ± 2 l/Mol was obtained for tert.-butyl hydroperoxide and 2,2,6,6-tetramethylpiperidine in ref. 64b). Further confirmation for an interaction between hindered amines and hydroperoxides is supplied by NMR measurements. Figure 10a shows part of the +OOH spectrum in toluene-d$_8$ (concentration 0.2 Mol/l) with the signal for the hydroperoxy proton at 6.7 ppm. Addition of as little as 0.002 Mol/l of tetramethylpiperidine to the same solution results in a displacement and marked broadening of the band (Fig. 10b). A similar observation was made with the amine II and its $\supset$N-methyl derivative.

From the standpoint of stabilization, complex formation is certainly an advantage. It means namely that the HALS stabilizer is already located preferentially at the site where degradation is initiated.

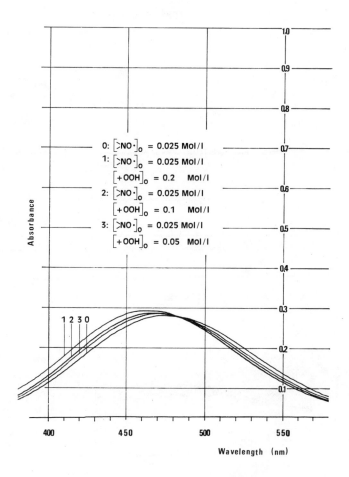

*Figure 9. Influence of the* tert-*butylhydroperoxide addition on the UV-absorption of the N—O· band (  NO·, I); solvent:* n-*heptane*

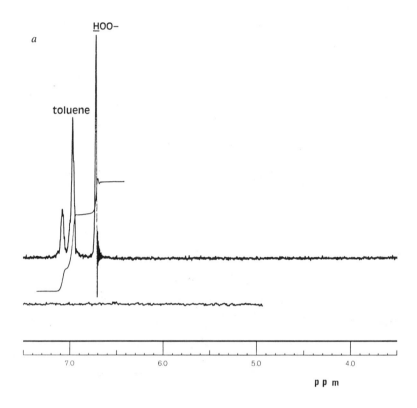

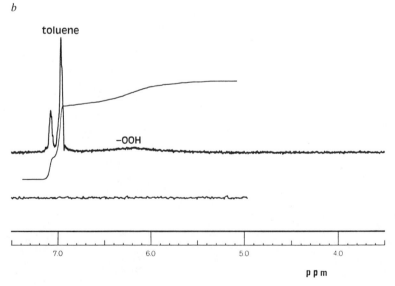

*Figure 10.    Part of the ¹H-NMR spectrum of (a)* tert-*butylhydroperoxide in tolu-ene-*d$_8$ *(0.2 mol/L) and (b)* tert-*butylhydroperoxide (0.2 mol/L) with added TMP (0.002 mol/L); solvent: toluene-*d$_8$

Amines and nitroxides in polymers

Tests on model substances in solution can throw much
light on certain aspects of the mode of action of additives. Experiments on polymers, on the other hand, are
more suitable for the phenomenological study of their
general effects. One such general effect is reflected
in the observation that during photooxidation of some
polymer systems containing amines the corresponding
nitroxides are usually formed, though in small amount
($\sim 10^{-4}$ Mol/l)[1,22,32].

During the last few years, the oxidation of hindered
amines to nitroxides has frequently been discussed in
the literature[21,22,32]. A convenient method of studying
this process represents the ESR spectroscopy. Polybutadiene with its high content of double bonds constitutes
on the other hand a readily oxidizable substrate which
allows a relatively fast investigation of oxidation
processes. Measurements of oxygen uptake as a function
of irradiation time made on 300 μm thick polybutadiene
sheets show (Fig. 11) that various N-substituted HALS
derivatives are also capable of protecting this polymer
from photooxidation. An interesting observation in this
connection is that the measured maximum $=$NO· concentration in the irradiated samples depends on the nature of
the N-substituent of the piperidine ring[20]. In sheets
containing 0.5 % ($\sim 10^{-2}$ Mol/kg) of amine II the maximum nitroxide concentration reached is of the order of
$10^{-4}$ Mol/kg. When the $=$N-CH$_3$ derivative is used the
value is only about $10^{-6}$ Mol/kg. For the $=$N-octyl derivative the nitroxide concentration is so small as to be
at the limit of sensitivity of the ESR technique.

The observed light-stabilizing effect of the $=$N-methyl
and $=$N-octyl derivatives (Fig. 11) - at least in the
case presented here - is thus manifested without formation of substantial amounts of nitroxide ($< 10^{-7}$ Mol/kg).


Conclusions

In the light not only of our own findings but also
those of other investigators, the following conclusions
as to the mode of action of amine light stabilizers can
be drawn:

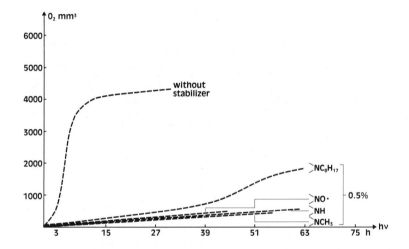

*Figure 11.*   *Oxygen uptake of 300 μm BR films (0.2 g) in air as a function of irradiation time in Xenotest 450; additives:* ⊃NH II *and its derivatives; conc. 0.5%*

- Since they are capable of intervening at various steps in the light-degradation process, TMP derivatives must be regarded as multifunctional UV stabilizers.

- One of the more important protective mechanisms is probably the ability of these substances to interact not only with various oxygen-centered radicals but also with hydroperoxides. This ability is supplemented by the formation of associates between the amine light stabilizer and species responsible for polymer degradation.

## Experimental

a) Photooxidation of diisopropyl ketone

Irradiations were carried out on an optical bench ($\lambda$ = 313 nm, I = 1.7 x $10^{-5}$ einstein/hr $cm^2$) in 1 cm quartz cells using light of a 200 W high pressure Hg lamp rendered parallel by passing through a lens. Nitroxide I or amine II was added to 5 x $10^{-2}$ Mol/l solutions of DiPK in benzene which were then flushed through with a continuous stream of oxygen (or nitrogen). The products were identified by GC, DC and MS using for comparison authentic samples. Structures of the two combination products between nitroxide I and isopropyl and isobutyryl radicals respectively were determined from the spectroscopic and analytical data.

Concentrations of educts and products were measured by gas chromatography using a Varian 2740 FID instrument (OV 101, Porapak columns).

b) Dimethylpentane oxidation

Experimental set up as in section a). Alkoxy radicals were produced by photolysis of di-tert.-butyl peroxide (2 Mol/l). The build up of hydroperoxide concentration was measured by a modified version of the iodometric method used by Carlsson and Wiles[68].

## Acknowledgement

The authors would like to thank the management of CIBA-GEIGY Ltd. in Basle cordially for permission to report the preceding results.

REFERENCES

1. Part IV, part III see B. Felder, R. Schumacher and F. Sitek, Helv. Chim. Acta, 63, 132 (1980).

2. R. Gächter and H. Müller, Kunststoff-Additive, Hanser, München, Wien 1979.

3. W.L. Hawkins, Ed., Polymer Stabilization, Wiley-Interscience, New York, 1972.

4. B. Ranby and J.F. Rabek, Photodegradation, Photo-oxidation and Photostabilization of Polymers, Wiley & Sons, London, 1975.

5. N.S. Allen and J.F. McKellar, British Polymer J. 9, 302 (1977).

6. J.F. McKellar and N.S. Allen, Photochemistry of Man-Made Polymers, Appl. Sci. Publishers, London, 1979.

7. S.L. Fitton, R.N. Howard and G.R. Williamson, British Polymer J. 2, 217 (1970).

8. H.J. Heller and H.R. Blattmann, Pure and Appl. Chem. 30, 145 (1972).

9. O. Cichetti, Adv. Polym. Sci. 7, 70 (1971).

10. D.J. Carlsson and D.M. Wiles, J. Macromol. Sci., Rev. Macromol. Chem. C 14, 155 (1976).

11. R.P.R. Ranaweera and G. Scott, Eur. Polym. J. 12, 825 (1976).

12. K.B. Chakraborty and G. Scott, Polym. Degrad. Stab. 1, 37 (1979).

13. F. Gugumus, Kunstst. Plast. 22, 11 (1975).

14. K.W. Leu, Plastic News, April, 10 (1975).

15. K.W. Leu, H. Linhart and H. Müller, Int. Conference on Polypropylene Fibers in Textiles, York, U.K. (1975) Plastics and Rubbers Publisher.

16. K. Berger, Kunststoffe Fortschrittsberichte 2, 79 (1976).

17. K.W. Leu, F. Gugumus, H. Linhart and A. Wieber, Plaste Kaut. 24, 408 (1977).

18. A.R. Patel and J.J. Usilton, Stabilization and Degradation of Polymers, Advances in Chem. Series 169, p. 116, ed. D.L. Allara and W.L. Hawkins, Amer. Chem. Soc. 1978.

19. A. Tozzi, G. Cantatore and F. Masina, Text. Res. J. 48, 433 (1978).

20. F. Sitek, 11th Conference of the Danube Countries on Natural and Artificial Aging of Plastics, Dubrovnik (Yugoslavia), October 1978.

21. V. Shlyapintokh, V. Ivanov, O. Khvostach, A. Shapiro and E. Rosantsev, Kunststoffe Fortschrittsberichte 2 (1), 25 (1976).

22. V. Shlyapintokh, V. Ivanov, O. Khvostach, A. Shapiro and E. Rosantsev, Dokl. Akad. Nauk SSSR 225, 1132 (1975).

23. M.N. Kusnietsova, L.G. Angert and A.B. Shapiro, Kauchuk i Resina 1977, 22.

24. K. Murayama, S. Morimura and T. Yoshioka, Bull. chem. Soc. Japan 42, 1640 (1969).

25. D. Bellus, H. Lind and J.F. Wyatt, J. Chem. Soc., Chem. Commun. 1972, 1199.

26. B. Felder and R. Schumacher, Angew. Makromol. Chem. 31, 35 (1973).

27. H.J. Heller and H.R. Blattmann, Pure and Appl. Chem. 36, 141 (1973).

28. J.B. Shilov and E.T. Denisov, Vysokomol. Soed. 16A, 2313 (1974).

29. V. Ivanov, S.Burkova, Ju. Morozov and V. Shlyapintokh, Vysokomol. Soed. 19B, 359 (1977).

30. K.B. Chakraborty and G. Scott, Chemistry & Ind. 1978, 237.

31. N.S. Allen and J.F. McKellar, J. Appl. Polym. Sci. 22, 3277 (1978).

32. D.J. Carlsson, D.W. Grattan and D.M. Wiles, Coatings and Plastics Preprints 39, 628 (1978).

33. D.J. Carlsson, D.W. Grattan, T. Suprunchuk and D.M. Wiles, J. Appl. Polym. Sci. 22, 2217 (1978).

34. D.W. Grattan, D.J. Carlsson and D.M. Wiles, Polym. Degrad. Stab. 1, 69 (1979).

35. K. Murayama, Farumashia (Japan) 10, 573 (1974).

36. B. Felder, R. Schumacher and F. Sitek, Chemistry & Ind. 1980, 155, 422.

37. F. Sitek, unpublished results, quenching of 2-pentanone photolysis in n-hexane.

38. N.S. Allen, J. Homer and J.F. McKellar, Makromol. Chem. 179, 1575 (1978).

39. N.S. Allen, J.F. McKellar and D. Wilson, Chemistry & Ind. 1978, 887.

40. N.S. Allen and J.F. McKellar, Chemistry & Ind. 1977, 537.

41. N.S. Allen and J.F. McKellar, Polym. Degrad. Stab. 1, 205 (1979).

42. K.B. Chakraborty and G. Scott, Polymer 18, 98 (1977).

43. D.M. Wiles, Pure and Appl. Chem. 50, 291 (1978).

44. J.E. Guillet, Pure and Appl. Chem. 36, 127 (1973).

45. F. Sitek, J.E. Guillet and M. Heskins, J. Polym. Sci., Symp. No. 57, 343 (1976).

46. J.E. Guillet, Stabilization and Degradation of Polymers, Advances in Chem. Series 169, p. 1, ed. D.L. Allara and W.L. Hawkins, Amer. chem. Soc. 1978.

47. J.E. Guillet, Prague Microsymposium on Macromolecules, 1979 p. L1.

48. N.J. Turro, J.Ch. Dalton, K. Dawes, G. Ferrington, R. Hautala, D. Morton, M. Niemczyk and N. Schore, Accounts Chem. Res. 5, 92 (1972).

49. P.J. Wagner, Accounts Chem. Res. 4, 168 (1971).

50. K. Schaffner and O. Jeger, Tetrahedron 30, 1891 (1974).

51. R.D. Small, Jr. and J.C. Scaiano, J. Am. Chem. Soc. 100, 4512 (1978).

52. N. Shimizu and P.D. Bartlett, J. Am. Chem. Soc. 98, 4193 (1976).

53. P.D. Bartlett and J. Becherer, Tetrahedron Letters 1978, 2983.

54. P.S. Engel, J. Am. Chem. Soc. 92, 6074 (1970).

55. W.K. Robbins and R.H. Eastman, J. Am. Chem. Soc. 92, 6076 (1970).

56. S.G. Whiteway and C.R. Masson, J. Am. Chem. Soc. 77, 1508 (1955).

57. K.U. Ingold in J. Kochi Ed., Free Radicals Vol. I, p. 66, Wiley, 1973, New York.

58. K.U. Ingold, Accounts Chem. Res. 2, 1 (1969).

59. N.A. Clinton, R.A. Kenley and T.G. Traylor, J. Am. Chem. Soc. 97, 3746 (1975); ibid 97, 3752 (1975); ibid 97, 3757 (1975).

60. M.B. Neiman and E.G. Rosantsev, Bull. Acad. Sci. SSSR, Chem. Ser. 1964, 1095.

61.  J.T. Brownlie and K.U. Ingold, Can. J. Chem. 45,
     2427 (1967).

62.  J.C. Chien, E.J. Vandenberg and H. Jabloner, J.
     Polym. Sci. A1, 381 (1968).

63.  M.U. Amin, G. Scott and L.M.K. Tillekeratne,
     Europ. Polym. J. 11, 85 (1975).

64a. J. Sedlar, J. Petruj and J. Pac, Prague Microsym-
     posium on Macromolecules, 1979, Poster M 47.

64b. J. Sedlar, J. Petruj, J. Pac and M. Navratil,
     Polymer 21, 5 (1980).

65.  D.W. Grattan, A.H. Reddoch, D.J. Carlsson and D.M.
     Wiles, J. Polym. Sci., Polym. Lett. Ed. 16, 143
     (1978).

66.  R. Hiatt in Organic Peroxides, Wiley-Interscience,
     N.Y., Vol. 2 p. 1, ed. D. Swern.

67.  N.V. Zolotova and E.T. Denisov, Bull. Acad. Sci.
     SSSR 1966, 736.

68.  D.J. Carlsson and D.M. Wiles, Macromolecules 2,
     597 (1969).

RECEIVED October 27, 1980.

# Differential Ultraviolet Spectroscopy as an Aid in Studying Polycarbonate Photodegradation

J. E. MOORE

General Electric Corporate Research and Development Center, Schenectady, NY 12301

The photochemical degradation of bisphenol-A polycarbonate (PC) has been the subject of a number of investigations.[1-9] As will be illustrated, our own studies on PC film weathering indicate that the major parameter which affects PC degradation is the amount of UV light to which the sample is exposed. This implicates photochemical reactions as principal degradation pathways.

Artificial weathering normally accelerates the natural weathering process by using a more intense source of UV light. Care must be taken, though, that high intensities of UV light at wavelengths much shorter than those found in natural sunlight (less than 295 nm) do not give misleading results.

A number of the previously cited investigators[1,2,5-9] have employed UV spectroscopy as an analytical tool for following PC degradation. We have found the measurement of UV spectra of weathered PC films by difference from an unexposed reference sample to be an extremely simple and useful analytical method. This nondestructive analysis allows the repetitive return of a sample to the exposure conditions and thus enables one to essentially perform continuous analyses on the same sample. This technique, of course, will not detect the formation of non-chromophoric products such as aliphatic oxidation products which may form during the degradation.

Previous work on PC photo-degradation mechanisms has mainly stressed the photo-Fries reaction which leads to the formation of substituted phenyl salicylates and dihydroxybenzophenones (Fig. 1). However, chain cleavage reactions which lead to phenolic products (Fig. 1) have also been reported to play important roles in PC weathering. The importance of oxygen on the photo-degradation of PC will be the subject of a forthcoming publication from this laboratory.[10] PC photo-degradation has been monitored using outdoor exposure at Schenectady, NY and under accelerated weathering conditions using RS sunlamps as a UV source.

0097-6156/81/0151-0097$05.00/0
© 1981 American Chemical Society

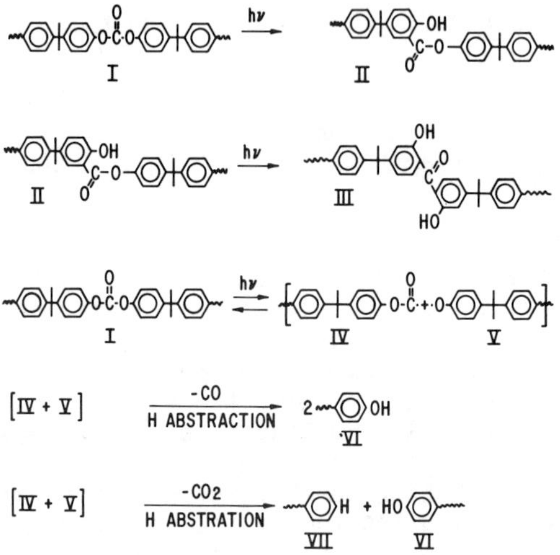

Figure 1.   Photo-Fries rearrangement and other photoinduced degradations of polycarbonate

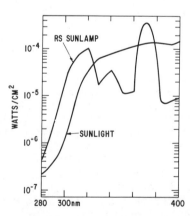

Figure 2.   Comparative light intensities: bright sunlight 2:15 p.m. 7/15/77 and 10 in. from RS sunlamp

## Experimental

Measurements of light intensity at various wavelengths, of different light sources were made with a Model 585-66 EG&G spectroradiometer. Plots of intensity vs. wavelength for natural sunlight and the RS sunlamp are shown in Figure 2.

Ten mil Lexan®*PC film containing no UV-stabilizers was obtained from Sheet Products Section, Mt. Vernon, Indiana. Reference films containing various additives were prepared by dissolving the additive and Lexan PC resin in dichloromethane and casting films from solutions containing about 5% solids. Portions of these films were selected which were 2 ± 0.1 mils thick.

UV spectra were recorded using a Perkin-Elmer Coleman 575 spectrometer. Yellowness index measurements (ASTM-D-1925) of film samples were made using a Colormaster Model V colorimeter.

Film samples were exposed to natural weathering conditions on an exposure rack constructed of wooden 2 X 4's. The samples were mounted directly on a 2 X 4 facing south at an angle of 45° to the horizontal. A measure of the UV light received by the samples was originally obtained from New York State Department of Environmental Conservation (ENCON) reports, kindly supplied by Mr. William Delaware of ENCON. These reports included both total sun and sky radiation and UV radiation measured by Eppley radiometers (photometers). The ENCON monitoring station in Schenectady is within four miles of the exposure rack so no appreciable differences in light intensity should be expected. More recently, a UV radiometer has been mounted directly on the exposure rack to obtain more accurate measurements.

Film samples were also exposed to UV light on RS sunlamp turntables. These consist of two RS sunlamps mounted 10 inches above 12 inch diameter turntables which were rotated at about 5 rpm. Alternate lamps were replaced every week.

## Results and Discussion

Natural Weathering. In order to follow chemical changes in PC resulting from weathering, ten mil PC films were examined by differential UV spectroscopy at intervals during natural and artificial weathering. The spectral changes were correlated with changes in the yellowness index of the samples and with conditions to which the samples had been exposed. Measurements of solar UV-exposure at Schenectady during 1977 from ENCON data are shown in Table 1.

---

* Lexan® is a registered trademark of the General Electric Company.

## TABLE I

### 1977 Outdoor Weathering Data from ENCON

| Latitude | 42"47'50" |
|----------|-----------|
| Elevation | 340 ft. |

$Calories/Cm^2$

|           | Total  | UV     | % UV |
|-----------|--------|--------|------|
| January   | 4154   | 175.2  | 4.07 |
| February  | 5096   | 222.0  | 4.36 |
| March     | 8432   | 381.6  | 4.53 |
| April     | 11280  | 456.0  | 4.04 |
| May       | 15500  | 629.2  | 4.06 |
| June      | 11820  | 531.0  | 4.49 |
| July      | 14911  | 558.2  | 3.74 |
| August    | 11470  | 462.0  | 4.03 |
| September | 7512   | 292.6  | 3.90 |
| October   | 5580   | 204.3  | 3.67 |
| November  | 3300   | 133.3  | 4.04 |
| December  | 2975   | 103.0  | 3.46 |
| 1977 TOTAL | 112030 | 4142.6 | 3.70 |

Typical differential UV spectra are shown in Figure 3. If the sample and reference films were exactly the same thickness, there would be no difference in absorbance at any wavelength, and only a straight line at absorbance = 0 would be obtained. In Figure 3, the sample film is slightly thicker than the reference film, so there is a small peak at about 285 nm. This is caused by the inability of the spectrometer to measure differences in PC absorbance when the absorbance is very high. However, when the absorbance is lower, as it is at longer wavelengths, the spectrometer is able to measure the difference in absorbances of 10 mil PC films starting at about 285 nm. Experience has shown that the disappearance of background noise indicates the wavelength cutoff for differential UV spectra. This wavelength corresponds to an absorbance of about 3-4 on a UV spectrum of a film versus air. During exposure to UV light, there is initially a slight decrease in the peak at about 285 nm, followed by a regular increase in this peak.

Simultaneously, there is the formation of a "negative peak" at about 310 nm in the early stages of exposure. A peak at about 340-360 nm also develops slowly during the exposure. The reasons for the initial decrease of the 285 nm peak and the formation of the negative peak at 310 nm are not completely clear. The phenomena are transitory but quite reproducable and real. An obvious explanation is that some chromophores in the exposed

sample are being destroyed or "bleached" during the exposure.  A
decrease in the yellowness index (YI) of the exposed sample is
also noted during this time.
    After the initial bleaching, there is a relatively rapid
increase in the peak at 287 nm and slower increases in the peaks
at 340-360 nm and 310-320 nm.  A plot of the increase in these
peaks as a function of UV dosage (cal/cm$^2$) is shown in Figure 4.
It is fairly certain that these peaks are due to phenolic, phenyl
salicylate, and dihydroxybenxophenone decomposition products
respectively as previously postulated.[1-3]  Figure 5 shows UV
spectra of 2 mil PC films containing one percent of these
ingredients.  It is apparent from the relative sizes of these
peaks and the corresponding peaks in spectra of degradated PC
that the major decomposition path for PC undergoing natural
weathering is a chain scission reaction leading to products
containing phenolic end groups.
    Evidence corroborating the formation of phenolic products in
weathered PC is shown in Figure 6.  In this figure, the differential infrared spectrum of a 10 mil PC film weathered for one
year at Schenectady shows a strong phenolic peak at 3500 cm$^{-1}$.
The identification of this peak is confirmed by the differential
IR spectrum of a 2 mil PC film containing 5% BPA shown in
Figure 7.  Intrinsic viscosity measurements of original and
weathered PC films indicate a decrease in the viscosity average
molecular weight ($M_v$) from 34,300 to 21,100 during the year of
weathering.  This confirms extensive degradation by chain
cleavage reactions, which lead to phenolic products, since the
photo-Fries rearrangement does not affect the polymer molecular
weight.
    The formation of the major UV degradation peak at about
287 nm in the weathered PC appears to correlate well with the
formation of the yellow color in the weathered sample.  In
Figure 8 the formation of both the peak at 287 nm and the yellow
color have been assumed to be products of a first order reaction.
This figure shows a plot of the log of the percentage of a
scaling constant minus the yellowness index divided by the
constant, versus a measurement of the exposure.  In this case,
the exposure is expressed as cal/cm$^2$, obtained from ENCON data.
It is apparent from Figure 8 that the formation of the 287 nm
peak and the yellow color are directly related and that both
apparently follow first order kinetics.

    Accelerated Weathering.  A number of instruments have been
devised for accelerating natural weathering conditions.  The
instruments usually incorporate a source of UV light since it is
the UV which rapidly degrades plastic samples.  One of the
simplest and cheapest sources of UV light is the GE RS sunlamp.
This has been used extensively for years to provide accelerating
weathering conditions.  The lamp consists of a medium pressure
mercury arc which is ballasted by the tungsten filament incan-

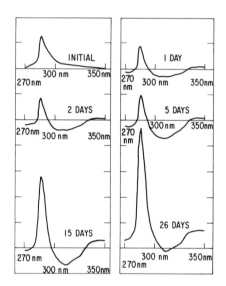

*Figure 3. Changes in PC UV spectra during natural weathering, June 1977*

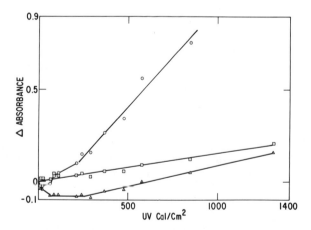

*Figure 4. Changes in PC UV spectrum peaks during natural weathering (( ⊙ ) 287 nm; ( △ ) 310–320 nm; ( ▣ ) 340–360 nm)*

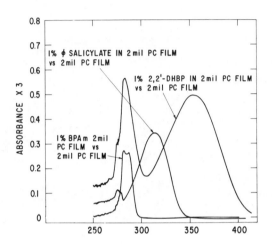

*Figure 5. Differential UV spectra of 2-mil PC films containing various additives*

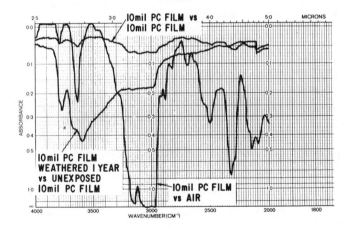

*Figure 6. IR spectra of 10-mil PC films*

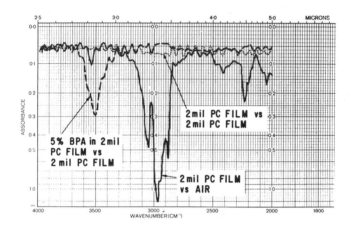

*Figure 7.   IR spectra of 2-mil PC films*

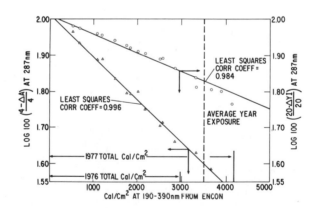

*Figure 8.   Formation of 287-nm peak and change in yellowness index in PC film during natural weathering*

descent portion of the lamp. A measurement of the UV output of
the RS sunlamp in comparison to natural UV light is shown in
Figure 2. A rotating exposure table was used to maximize uni-
formity of sample exposure under the RS sunlamps. Typical
differential UV spectra of 10 mil PC films exposed under RS
sunlamps are shown in Figure 9. Again the major spectral changes
caused by the UV light exposure are the formation of a sharp peak
at about 287 nm and a broader peak at about 320 nm. The formation of the peak at 320 nm is in contrast to the "bleaching" in
this area in samples exposed to natural light (compare Figures 3
and 7). The formation of the peak at about 320 nm is related to
the light intensity at wavelengths below about 300 nm. As shown
in Figure 2 the light intensity at 300 nm is about 20 times
higher under the RS sunlamp than in natural sunlight.

  Yellowness indices of the exposed PC samples are easily
measured in the colorimeter during the course of the exposure. A
plot of the increases in absorption at 287 nm and in yellowness
index ($\Delta YI$) versus exposure time under RS sunlamps is shown in
Figure 10 (as was shown for the same parameters with natural
sunlight exposure in Figure 8). From these two plots, the points
where the changes in absorption at 287 nm and in yellowness index
are equal can be determined and compared with the exposures
needed to effect these changes. This comparison indicates that
one year of natural exposure at Schenectady is equivalent to 170
hours of RS sunlamp exposure when changes in absorption at 287 nm
are measured and to 140 hours of RS sunlamp exposure when changes
in yellowness index are determined.

## Conclusions

  1) The use of differential UV spectroscopy is a facile
analytical tool, providing a rapid, non-destructive method for
determining the course and extent of degradation of PC films
during accelerated or natural weathering.
  2) Chain cleavage with subsequent formation of phenolic
products, rather than the photo-Fries rearrangement to form
salicylates and dihydroxybenxophenones, has been identified as
the major initial degradation pathway of PC exposed to natural
weathering conditions.
  3) The spectroscopic method is also useful for correlating
the rate of PC degradation under accelerated weathering to that
under natural weathering conditions.

## Acknowledgements

  The author thanks the General Electric Research and
Development Center for permission to publish this work and thanks
Mr. S. T. Rice for many of the yellowness index and UV spectral
measurements. Mr. William Delaware of the New York State Department of Environmental Conservation is also thanked for his data.

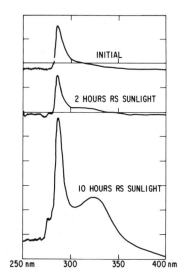

*Figure 9. UV spectra of PC films exposed under RS sunlamps*

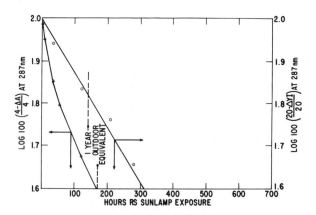

*Figure 10.   Formation of 287-nm peak and change in yellowness index in PC film during exposure under RS sunlamps*

## Abstract

Films of bisphenol-A polycarbonate were weathered on an outdoor rack in Schenectady, NY and under accelerated conditions under RS sunlamps. Changes in the samples were measured by changes in yellowness index (ASTM-D-1925) and by changes in the UV spectra relative to an unexposed reference. The formation of phenolic degradation products followed by formation of photo-Fries products (salicylates and dihydroxybenxophenones) can readily be detected in the UV spectra. The relative amounts of phenolic or photo-Fries products are greatly influenced by the wavelength of light used for the photo-degradation. The use of short wavelength UV light (<300 nm) leads to larger amounts of photo-Fries products. The formation of phenolic degradation products (UV peak at about 287 nm) appears to correlate well with increases in yellowness index of the polycarbonate. Both the formation of the 287 nm peak and the increase in yellowness are related to the UV dosage received by the sample as measured by hours of exposure to the RS sunlamps or by UV Langleys as measured on the roof rack by an Eppley radiometer.

## Literature Cited

1. D. Bellus, P. Hrdlovic and Z. Manasek, Polymer Letters, 4, 1 (1966).
2. S. Tahara, Chem. High Polymers 23 (253), 303 (1966) Japan.
3. A. Davis and J. H. Golden, J. Macromol. Sci.-Revs. Macromol. Chem., C3 (1) 49 (1969).
4. B. D. Gesner and P. G. Kelleher, J. Appl. Poly. Sci. 13, 2183 (1969).
5. P. A. Mullen and N. Z. Searle, J. Appl. Poly. Sci. 14, 765 (1970).
6. J. S. Humphrey and R. S. Roller, Mol. Photochem. 3 (1), 35 (1971).
7. J. S. Humphrey, A. R. Shultz and D. B. G. Jaquiss, Macromolecules 6, 305 (1973).
8. A. Gupta, A. Rembaum and J. Moacanin, Macromolecules, 11 (6), 1285 (1978).
9. E. Ong. and H. E. Bair, Polymer Preprints 20 (1), 945 (1979).
10. A. Factor and M. L. Chu, Polymer Degradation and Stability, in press.

RECEIVED September 16, 1980.

# Photostabilization of Bisphenol A-Epichlorohydrin Condensation Polymers
## Fluorescence and Model Compound Studies

S. PETER PAPPAS, LESLIE R. GATECHAIR, ELLIS L. BRESKMAN, and RICHARD M. FISCHER

Polymers and Coatings Department, North Dakota State University, Fargo, ND 58105

U. K. A. KLEIN

Institute for Physical Chemistry, University of Stuttgart, Pfaffenwaldring 55, 7000 Stuttgart 80, West Germany

Previous studies have provided evidence for resonance energy transfer (RET) from excited singlet states of aromatic polymers 1 and 2 to 2-hydroxybenzophenone light stabilizers.[1] The evidence was derived from fluorescence intensity measurements on polymer films as a function of stabilizer concentration. After correcting for absorption of exciting and emitted light by the stabilizer (screening and radiative energy transfer, respectively),[2] the resulting experimental quenching efficiencies exhibited good agreement with predicted RET quenching efficiencies, based on Förster kinetics.[3]

During the course of these studies, it was found that fluorescence intensity from the polymeric films rapidly decreased on continued excitation in a fluorescence spectrophotometer (ca. 30% loss in 1 min for 1). Herein, we (1) elaborate further upon the fluorescence loss studies, (2) provide direct evidence for RET from fluorescence lifetime measurements, and (3) present preliminary findings on the photochemistry of model compounds for polymer 1. The results support the conclusion, from previous studies, that the effectiveness of added stabilizer decreases with time due to formation of a photoproduct(s) from the polymer which competes in RET, and is less able to dissipate the resulting excitation energy.[1]

## Results and Discussion

Fluorescence Intensities. The fluorescence from polymeric films of bisphenol A-epichlorohydrin condensate 1 (Eponol-55-B-40,

Shell Chemical Company), exhibits a maximum at 300 nm, corresponding to that of the model chromophore anisole. The fluorescence intensity decreases monotonically with increasing concentration of 2,4-dihydroxybenzophenone (DHB) and, furthermore, decreases with time on continued excitation (274 nm) in the spectrophotometer. The fluorescence loss with time may be resolved into two exponential decays. Initially, a relatively rapid fluorescence loss is observed within 20 sec, followed by a slower loss. Loss constants for the initial ($k_1$) and secondary ($k_2$) exponential decays for 1.5 μm films (on glass slides) containing varying concentrations of DHB are provided in Table I (entries 1-3). The initial loss constants are seen to decrease more markedly with increasing DHB concentration than the secondary constants.

In order to determine the effect of air on fluorescence loss, free films of polymer 1 (15 μm thick) were placed in a quartz cuvette, which was evacuated prior to excitation in the fluorescence spectrophotometer. Although the initial loss constant was not determined accurately, both constants (entry 4) were substantially smaller in vacuo relative to air. Fluorescence loss from correspondingly thick films in air is provided in entry 5.

TABLE I.
Fluorescence Intensity Loss Constants[a] from Polymer 1 Films

| DHB $\underline{M} \times 10^3$ | Film Thickness (μm) | $k_1 \times 10^3 (sec^{-1})$ | $k_2 \times 10^3 (sec^{-1})$ |
|---|---|---|---|
| 1.   0 | 1.5 | 9.9 | 2.0 |
| 2.   2.3 | 1.5 | 4.6 | 1.7 |
| 3.   9.2 | 1.5 | 2.8 | 1.5 |
| 4.[b]  0 | 15 | <1 | 0.18 |
| 5.   0 | 15 | 6.9 | 1.2 |

[a]Initial ($k_1$), secondary ($k_2$), average of 2 determinations, estimated error ± 15%.   [b] In vacuo.

Fluorescence Lifetimes. Fluorescence lifetimes were determined by the phase shift method,[4] utilizing a previously-described phase fluorimeter.[5] The emission from an argon laser was frequency doubled to provide a 257 nm band for excitation. Fluorescence lifetimes of anisole and polymer 1 in dichloromethane solution were 2.2 and 1.4 nsec, respectively. Fluorescence lifetimes of polymer 1 films decreased monotonically with increasing DHB concentration from 1.8 (0) to 0.7 nsec (9.2 x $10^{-3}$ $\underline{M}$DHB). Since fluorescence lifetimes (in contrast to fluorescence intensities) are unaffected by absorption effects of the stabilizer, these results provide direct evidence in support of the intensity measurements[1] for RET from polymer to stabilizer.

On continued excitation in the phase fluorimeter, the fluorescence lifetime of polymer 1 films also decreased with time. The lifetime decrease was exponential with an average loss constant of $8.2 \pm 1.2 \times 10^{-4}$ sec$^{-1}$ (1.5 μm thick film) from measurements at different sites on the film. These findings constitute direct evidence for RET from the polymer to a photoproduct(s) in support of the fluorescence intensity measurements.

Significance of Photoproduct Formation. Evidence for photoproduct formation had also been obtained from previous attempts to correlate the efficiency of RET quenching with stabilization of the films against discoloration in an accelerometer (>280 nm).[1] For this purpose, stabilization efficiency was defined as $1-A_s/A_0$, where $A_s$ and $A_0$ represent the increase in absorbance in the blue spectral region (yellowing) in the presence and absence of stabilizer, respectively. The resulting stabilization efficiencies were found to decrease substantially over relatively short exposure times (ca. 40% decrease between 10 and 25 hrs irradiation). Difference absorption spectra obtained during accelerometer exposure exhibited a new absorption band at ca. 300 nm which overlapped strongly with polymer fluorescence (required for efficient RET quenching) and weakly with polymer absorption (screening).[1]

With increasing concentration of DHB, the photoproduct forms more slowly, as evidenced by decreasing loss of fluorescence intensity (Table I, entries 1-3). Nevertheless, the concentration of photoproduct(s) and RET from the polymer to photoproduct(s) are expected to increase with time, and stabilization of the polymer will eventually depend upon the capability of the photoproduct(s) to dissipate excitation energy imparted in the RET process. The observed decrease in stabilization efficiency by DHB (based on film discoloration) with exposure time in an accelerometer indicates that DHB is more effective than the photoproduct(s) in dissipating the light energy. Similar spectroscopic studies on polystyrene have led to the same conclusion in this case, as well.[6]

The capability of 2-hydroxybenzophenone derivatives to dissipate light energy has been ascribed to rapid deactivation of the excited singlet state by intramolecular interaction between the carbonyl and hydroxyl groups, possibly involving reversible H-transfer.[7] These proposals are outlined in Scheme I, where P and PP represent the polymer and photoproduct, respectively.

Photochemistry of Model Compounds. Preliminary photochemical studies have been carried out on 1,3-diphenoxy-2-propanol (3)[8] as a model compound for bisphenol A-epichlorohydrin condensates 1. The utilization of 3 as a model compound for thermal degradation of 1 has been reported.[9] Irradiation (254 nm) of 3 in acetonitrile (N$_2$ purge) provides two major volatile products, which have been identified as phenol and phenoxyacetone (4), by comparison of retention times (gas chromatography) with known samples. A possible mechanism for

SCHEME I

Absorption:          $P \xrightarrow{h\nu} {}^1P*$

Fluorescence:       ${}^1P* \longrightarrow P + h\nu'$

Photoproduct formation:  ${}^1P* \longrightarrow PP$

RET to stabilizer:

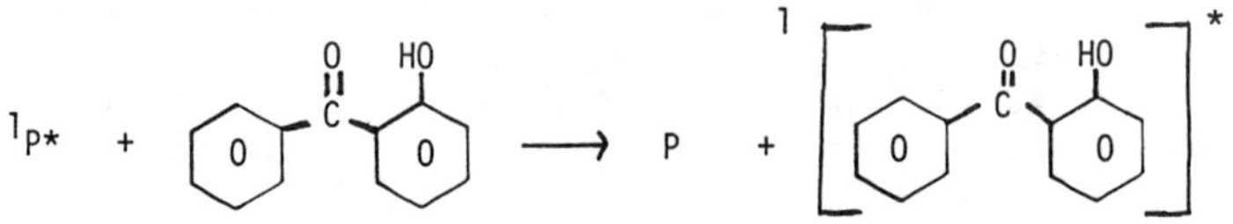

Energy dissipation (light⟶chemical⟶thermal energy):

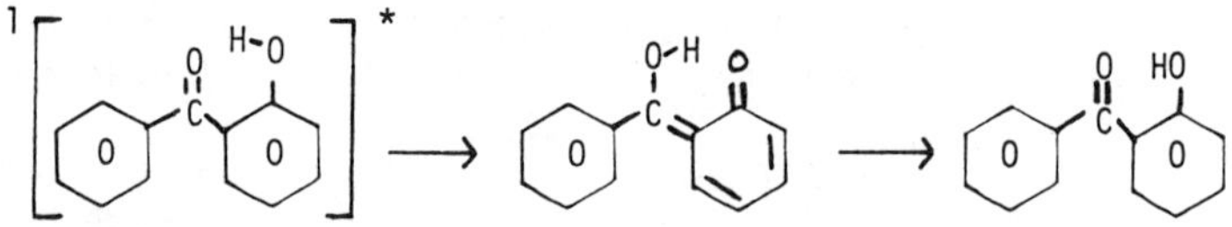

RET to photoproduct:  ${}^1P* + PP \longrightarrow {}^1PP* + P$

${}^1PP* \longrightarrow$ Photodegradation

SCHEME II

formation of these products is provided in Scheme II. Initial C-O bond homolysis has been postulated as the primary step in the photochemistry of aryl esters and ethers, based on flash photolysis studies.[10] Subsequent H-transfer (which may occur within a solvent cage) followed by ketonization of the resulting enol yields the products.

The importance of phenol formation by the proposed pathway was probed by irradiating 1,3-diphenoxy-2-methyl-2-propanol (5) under the same conditions. Compared to 3, the rate of phenol formation was approximately 2 times slower. Since the H-transfer step in Scheme II is not available to 5, the results provide support for the scheme as an important, but not sole, pathway for phenol formation. Irradiation of 3 and 5 with an air purge resulted in faster rates of phenol formation (ca. 5-fold) relative to $N_2$. These findings parallel the accelerated fluorescence intensity loss from polymer 1 films in air as compared to the results in vacuo (see Table I).

$$Ph-O-CH_2-\underset{\underset{CH_3}{|}}{\overset{\overset{OH}{|}}{C}}-CH_2-O-Ph$$

5

Irradiation of 3 at longer wavelengths (>280 nm) provided phenyl formate (6) as a major volatile product, together with minor amounts of phenol and phenoxyacetone (4), as well as other products. A possible pathway for formation of phenyl formate by oxidation and subsequent cleavage is provided in Scheme III. Phenoxyacetic acid (7) was also identified as a minor product by mass-gc analysis. Photolysis of phenoxyacetone (4)[11] and phenoxyacetic acid (7)[12] yields phenol together with photo-Fries products (also shown in Scheme III).

At present, the relevance of these results to photodegradation[13] of condensates 1 is a matter of speculation. Of particular interest is identification of the photoproduct quencher(s) (PP, Scheme I). Possible candidates are salicylic acid derivatives, which exhibit the requisite absorptivity at about 300 nm, and which may be formed by oxidation of ortho-photo-Fries products (Scheme III), as illustrated in eq. 1.

(1)

SCHEME III

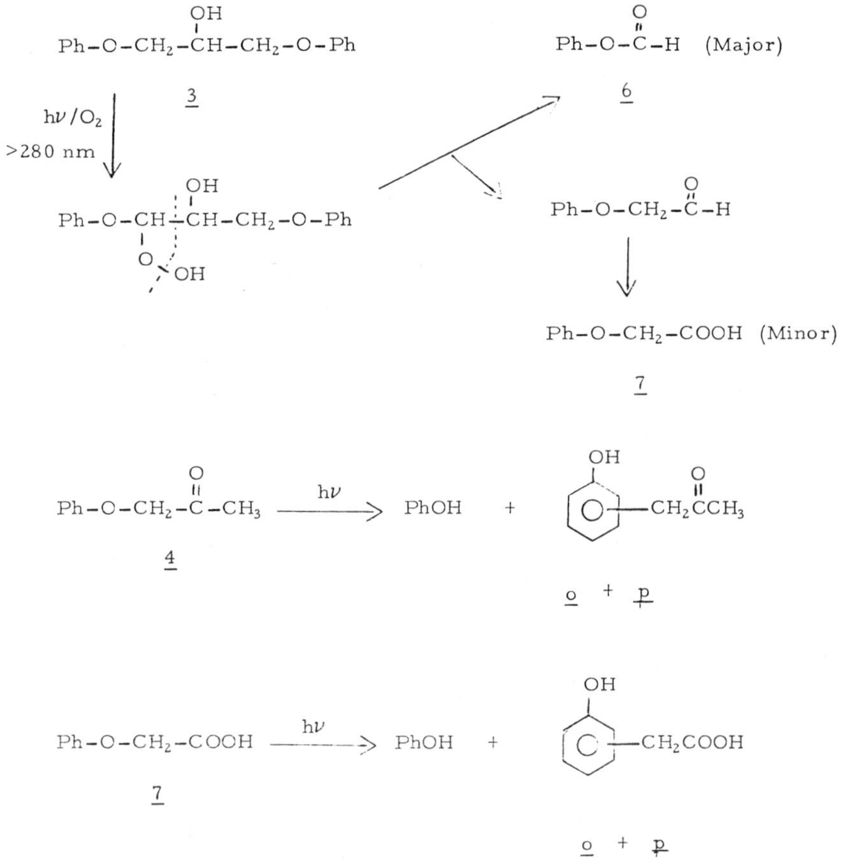

$$
\underset{\underline{3}}{\text{Ph}-\text{O}-\text{CH}_2-\overset{\overset{\text{OH}}{|}}{\text{CH}}-\text{CH}_2-\text{O}-\text{Ph}}
\qquad\qquad
\underset{\underline{6}}{\text{Ph}-\text{O}-\overset{\overset{\text{O}}{\|}}{\text{C}}-\text{H}} \quad (\text{Major})
$$

$\text{h}\nu / \text{O}_2$
$>280$ nm

$$
\text{Ph}-\text{O}-\text{CH}\overset{\overset{\text{OH}}{|}}{—}\text{CH}-\text{CH}_2-\text{O}-\text{Ph}
\qquad
\text{Ph}-\text{O}-\text{CH}_2-\overset{\overset{\text{O}}{\|}}{\text{C}}-\text{H}
$$
$$
\overset{|}{\text{O}}{\underset{\text{OH}}{\diagdown}}
$$

$$
\underset{\underline{7}}{\text{Ph}-\text{O}-\text{CH}_2-\text{COOH}} \quad (\text{Minor})
$$

$$
\underset{\underline{4}}{\text{Ph}-\text{O}-\text{CH}_2-\overset{\overset{\text{O}}{\|}}{\text{C}}-\text{CH}_3}
\xrightarrow{\text{h}\nu}
\text{PhOH} \quad + \quad
$$

$$\underline{o} \;+\; \underline{p}$$

$$
\underset{\underline{7}}{\text{Ph}-\text{O}-\text{CH}_2-\text{COOH}}
\xrightarrow{\text{h}\nu}
\text{PhOH} \quad + \quad
$$

$$\underline{o} \;+\; \underline{p}$$

Experimental

Materials. Bisphenol A-epichlorohydrin condensate 1
(Eponol-55-B-40, Shell Chemical Co.) was precipitated from
chloroform solution by addition of methanol three successive
times prior to utilization.  Polymer films of 1.5 and 15 μm were
cast onto glass plates from chloroform solution.
2,4-Dihydroxybenzophenone was obtained from Aldrich Chemical
Co. and used as received.
1,3-Diphenoxy-2-propanol (3) was prepared from phenol and
epichlorohydrin (1-chloro-2,3-epoxypropane), as previously
described,[8] and recrystallized three times from 2-propanol to
yield white crystals, m.p. 82-82.5°C.  The nmr spectrum in
CDCl$_3$ (Varian EM-390 spectrometer) exhibited resonances (in
ppm (δ) relative to tetramethylsilane) at 3.0 (1H, doublet,
J=5 Hz), 4.1 (4H, doublet, J=5 Hz), 4.3 (1H, multiplet, J=5 Hz),
and 6.8-7.4 (10H, multiplet), which are assigned to the hydroxyl,
methylene, methyne, and aryl hydrogens, respectively.
1,3-Diphenoxy-2-methyl-2-propanol (5) was prepared from
phenol and 2-methylepichlorohydrin (1-chloro-2-methyl-2,3-
epoxypropane) by the above method[8] and obtained as an oil,
which exhibited a single peak on gas chromatographic (gc)
analysis.  In conformance with the proposed structure, the nmr
spectrum in CDCl$_3$ exhibited a multiplet at 6.8-7.4 ppm (aromatic
hydrogens), and singlets at 4.0, 3.1 and 1.4 ppm, corresponding
to the methylene, hydroxyl and methyl hydrogens, respectively.
2-Methylepichlorohydrin[14] was obtained from epoxidation of
methallylchloride with meta-perbenzoic acid, by a standard
procedure.[15]

Fluorescence Studies.  Fluorescence spectra of films on
glass plates were obtained with a Perkin-Elmer MPF-3 spectro-
fluorimeter.  A previously-described phase fluorimeter[5] was
utilized for fluorescence lifetime determinations.

Irradiation Studies.  Irradiation of 1,3-diphenoxy-2-
propanol (3), 0.01 M in acetonitrile, was conducted at 254 nm,
utilizing a 2.5-W low pressure Hg immersion lamp (PCQ9G-1,
Ultraviolet Products), and at wavelengths longer than 280 nm,
utilizing a Hanovia 450-W high-pressure Hg immersion lamp
(Type L) and 9700 Corex filter sleeve.  Irradiation of 1,3-
diphenoxy-2-methyl-2-propanol (5), 0.01 M in acetonitrile, was
also conducted at 254 nm.
The photolysis vessels were equipped with a gas inlet,
serum-capped opening for aliquot removal, and a water-cooled
condenser.  The irradiated solutions also contained decalin
(1.5 x 10$^{-3}$ M), which was utilized as an internal standard for
gc analysis (Varian Aerograph 2400, flame ionization detector,
6' x 1/8" columns of OV-101 (1.5%) on Chromosorb G).  Product
analysis was also conducted with a Varian mass spectrometer

(112-8) interfaced with a gc (3700) and data collection system (SS 200).

## Acknowledgement

We are grateful to Professors M. Hauser and H. E. A. Kramer for stimulating discussions, the Deutscher Akademischer Austauschdienst (DAAD) for a research grant (to SPP), DeSoto, Inc. for financial assistance, and Shell Chemical Company for generously supplying resins.

## Literature Cited

1.  Breskman, E.L.; Pappas, S.P.  J. Coatings Technol., 1976, 48 (622), 34.

2.  Breskman, E.L., Ph.D.  Thesis, North Dakota State University, 1976.

3.  Förster, Th., Discuss. Faraday Soc., 1959, 27, 7; North, A.M.; Treadaway, M.F., Eur. Polymer J., 1973, 9, 609.

4.  Ware, W.R., "Creation and Detection of the Excited State," Lamola, A.A., Ed., Vol. I, Part A, 1971, pp. 269-283.

5.  Haar, H-P.; Hauser, M.  Rev. Sci. Instrum., 1978, 49, 632.

6.  Pivovarov, A.P.; Pivovarova, T.S.; Lukovnikov, A.F.  Polymer Sci. (USSR), 1973, 15, 747.

7.  Klöpffer, W.  Adv. Photochem., 1977, 10, 311

8.  Minor, W.F.; Smith, R.R.; Cheney, L.C.  J.Amer.Chem.Soc., 1954, 76, 2993.

9.  Paterson-Jones, J.C.; Percy, V.A.; Giles, R.G.F.; Stephen, A.M.  J. Appl. Polym. Sci., 1973, 17, 1877.

10.  Kalmus, C.E.; Hercules, D.M.  J.Amer.Chem.Soc., 1974, 96, 449.

11.  Dirania, M.K.M.; Hill, J.  J.Chem.Soc.(C), 1968, 1311

12.  Kelly, D.P.; Pinhey, J.T.  Tetrahedron Lett., 1964, 3427.

13.  Kelleher, P.G.; Gesner, B.D.  J.Appl. Polym.Sci., 1969, 13, 9; Gesner, B.D.; Kelleher, P.G., ibid., 1969, 13, 2183.

14.  DePuy, C.H.; Dappen, G.M.; Eilers, K.L.; Klein, R.A.  J.Org. Chem., 1964, 29, 2813.

15.  Fieser, L.F.; Fieser, M.  "Reagents for Organic Synthesis," Vol. I, John Wiley and Sons, Inc., 1968, pp. 135-6.

RECEIVED September 16, 1980.

# 9

# Photochemistry of *N*-Arylcarbamates

CHARLES E. HOYLE, THOMAS B. GARRETT, and JOHN E. HERWEH

Armstrong World Industries, Inc., Research and Development Center, 2500 Columbia Avenue, P.O. Box 3511, Lancaster, PA 17604

The use of isocyanates in coatings' formulations has had and will continue to have an important role in providing durable finishes. Due primarily to the fact that polyurethanes based upon aromatic diisocyanates undergo photodegradation and accompanying discoloration, the utilization of the considerably more costly aliphatic diisocyanates has been mandated. Urethanes derived from the latter have found widespread use in coatings, in spite of the fact that they also apparently undergo photodegradation. Typically, however, discoloration is not associated with their photodegradation.

The mechanism involving the photochemical-induced degradation of urethanes derived from aromatic diisocyanates has remained somewhat of an enigma. The lack of a thorough understanding of the elements leading to degradation and ultimate discoloration of urethanes based upon aromatic diisocyanates has detracted from efforts to find suitable means to provide lasting stabilization.

A number of studies have dealt with the photo-induced discoloration and degradation of polyurethanes based on aromatic diisocyanates such as toluene diisocyanate (TDI - represents a mixture of 2,4-toluene diisocyanate and 2,6-toluene diisocyanate isomers) and methylene 4,4-diphenyl diisocyanate (MDI) (1, 2, 3, 4, 5). It has been suggested that the MDI and TDI based polyurethanes photodegrade (Scheme I) by a photo-Fries rearrangement

Scheme I (Rearrangement Products)

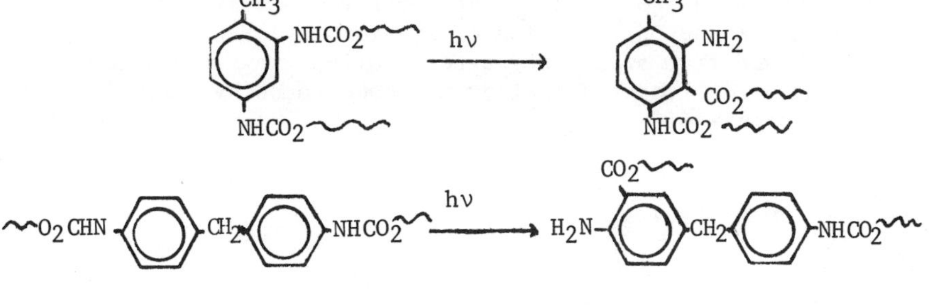


0097-6156/81/0151-0117$05.00/0

process (4). Formation of quinone imide (quinone methine imide) products have also been postulated (Scheme II) (1, 2, 3). In order to better understand the photoprocesses of actual polyurethane coatings based on MDI or TDI, researchers have studied the photochemistry of ethyl N-phenylcarbamate (1a) as a model

Scheme II (Quinone Imide and Quinone Methine Imide Products)

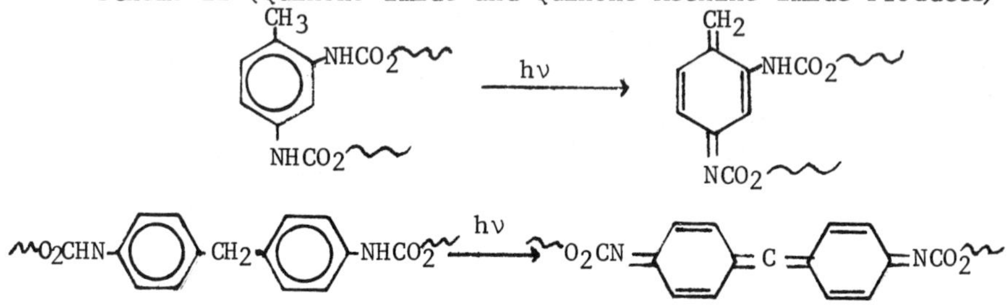

system (6-11). It was reported that the photolysis of 1a is zero order at low conversions and self-inhibiting at higher conversions due to interference by absorbing products (6). Schwetlick and co-workers (8) found that irradiation of 1a at 254 nm yielded as major identifiable products aniline (1b), ethyl o-aminobenzoate (1c), and ethyl p-aminobenzoate (1d). They proposed that the products were formed predominantly by N-C bond cleavage resulting in a solvent-caged radical pair. Within the solvent cage the ethoxycarbonyl radical attacked the phenyl ring at the ortho and para positions to give the reported photo-Fries products. Similarly, aniline (1b) was formed by diffusion of the anilinyl radical from the solvent cage followed by hydrogen abstraction.

In actual polyurethane coatings based on TDI there is a methyl group ortho or para to the reactive carbamate ($-NHCO_2R$) group. Thus, since 1a has no methyl groups ortho or para to the carbamate group, it is questionable whether it is an appropriate model system for polyurethanes based on TDI. A better model for the arylcarbamate moiety ($-ArNHCO_2R-$) in a polyurethane based on TDI (or MDI) would have methyl groups ortho and para to the reactive carbamate group. It is thought that substitution of methyl groups on the phenyl ring might alter the reactivity of the radicals formed upon photolysis. The current investigation is directed toward the photodegradation of simple alkyl N-arylcarbamates 2a-4a derived from aryl isocyanates bearing ring-substituted methyl groups. Products similar to those found by Schwetlick (8) upon photolysis of 1a are expected.

la  $R_1$ = Et; $R_2$, $R_3$, $R_4$ = H

2a  $R_1$ = Pr; $R_2$ = $CH_3$; $R_3$, $R_4$ = H

3a  $R_1$ = Pr; $R_3$ = $CH_3$; $R_2$, $R_4$ = H

4a  $R_1$ = Pr; $R_2$, $R_3$, $R_4$ = $CH_3$

1b  $R_2$, $R_3$, $R_4$ = H

2b  $R_2$ = $CH_3$; $R_3$, $R_4$ = H

3b  $R_3$ = $CH_3$; $R_2$, $R_4$ = H

4b  $R_2$ = $R_3$ = $R_4$ = $CH_3$

lc  $R_1$ = Et; $R_3$, $R_4$ = H

2c  $R_1$ = Pr; $R_3$ = H; $R_4$ = $CH_3$

3c  $R_1$ = Pr; $R_3$ = $CH_3$; $R_4$ = H

1d  $R_1$ = Et; $R_2$, $R_4$ = H

2d  $R_1$ = Pr; $R_2$ = H; $R_4$ = $CH_3$

## Experimental

Material Preparation. The arylamines (Aldrich Chemical Company) 1b, 2b, 3b, and 4b were either distilled or sublimed before use. The alkyl N-arylcarbamates (1a-4a) and the bispropyl carbamate of 2,4-TDI (5a) were prepared according to the following general procedure. Dry propanol (100% molar excess) and a catalytic (1% by wt. of isocyanate) amount of pyridine were placed in a flame-dried flask under a nitrogen atmosphere. A solution of the requisite isocyanate in ethyl acetate (dry) was added dropwise with stirring to the alcohol/pyridine solution. The extent of reaction was determined by following the intensity of the isocyanate band (ca. 2270-2240 $cm^{-1}$) in the IR. When the isocyanate had completely reacted, the cooled reaction mixture was filtered to remove insolubles. The filtrate was concentrated at reduced pressure and purified by appropriate means. The structures assigned to the various carbamates were confirmed by NMR spectroscopy. The amino and methyl substituted benzoates (1c, 1d, 2c, 2d, and 3c) were prepared by esterification of their corresponding substituted benzoic acids using boron trifluoride etherate as catalyst. The crude propyl benzoates were purified by fractional distillation or recrystallization from hexane. Structural assignments were confirmed by NMR spectroscopy. Elemental analysis was obtained for all new compounds synthesized.

Solution Photolysis. Solutions were prepared by dissolving the appropriate carbamate (1a-4a) or amine in cyclohexane (spectrograde –

Burdick and Jackson).  All solutions were photolyzed to less than
5% conversion in a standard 3 ml capacity, 1-cm path length quartz
cell.  Samples were irradiated with a 450-Watt medium pressure,
Hanovia mercury lamp focused through an appropriate band-pass
filter (280 nm or 254 nm) onto the 1-cm quartz cell with the requisite solution.  Test solutions could be purged with either
helium or oxygen using a needle valve assembly attached to the
tapered quartz cell neck.  The loss of carbamate due to photolysis
and the amounts of known photoproducts were determined quantitatively by GC using eicosane as an internal standard.  The columns
were 6' stainless steel containing Carbowax 20M on chromosorb G.
A ferrioxalate actinometer was used to determine the lamp light
intensity (12).  The quantum yield of loss ($\Phi_D$) and of product
formation ($\Phi_P$) were then calculated by standard methods (12).
The biscarbamate 5a and carbamate 3a were dissolved in acetonitrile (Fischer – ACS grade) and photolyzed with a 200-Watt
medium pressure, Hanovia lamp in a typical preparative photolysis
apparatus with pyrex sleeve.  Integrated proton NMR data for the
resultant solutions were made on a Jeol 4H-100 NMR.

     Film Photolysis.  Inhibitor free methyl and propyl methacrylate
(Polysciences, Incorporated) were then bulk polymerized under a
He atmosphere in sealed tubes using AIBN (0.1% by wt.) as an
initiator.  Typically polymerization was effected by heating @ 70°C
– ca. 4 hours for methyl methacrylate and 6 hours for propyl
methacrylate.  Polymethylmethacrylate and polypropylmethacrylate
lacquers (8% by wt. of polymer) were prepared in 1,2-dichloro-
ethane.  The carbamates or photodegradation products (up to 10%
by wt. of polymer) were added to aliquot portions of the polymer
lacquers.  The resulting lacquers were applied to glass plates
using a 6 mil Bird film applicator.  The drawdowns were air-dried
(3-1/2 hrs) and then dried in vacuo at rt for 16 hrs.  The films
(ca. 1-1/2 mils) were removed from the glass plates by immersion
in water.  The free films were dried in vacuo (<1 mm) at rt in
the presence of $P_2O_5$ for ca. 15 hrs.  The dry films containing the
carbamates were irradiated using the same apparatus described
above with the 280 nm band pass filter.  The light falling on the
polymer film per unit area was calculated using ferrioxalate
actinometry.  A portion (weight determined) of the photolyzed or
unphotolyzed PMMA or PPMA film containing the requisite carbamate
was dissolved in THF.  The resulting solution containing a known
amount of sulfolane, as an internal standard, was diluted to 10
ml.  Samples of the resulting solution were used for GC analysis.
A liner containing glass wool was installed in the injection port
of the GC to trap polymer residues.  A solution containing a known
amount of the carbamate in THF along with sulfolane as an internal
standard was used to establish the concentration of carbamates in
the PMMA and PPMA matrix.  Quantum yields were then determined.
Product ratios were calculated from UV difference spectra (taken on
a Beckman DK-2A spectrometer) of films before and after photolysis.

M.O. Calculations. The semi-empirical molecular orbital calculations were made using the UHF INDO model developed by Pople
and co-workers (13), which incorporates the one-center exchange
integral. Additionally, instead of assuming standard values for
bond distances and angles, full geometry optimization at the INDO
level was employed (14). Thus the results do not depend upon an
arbitrary choice for the molecular geometry.

Results and Discussion

Disappearance Quantum Yields ($\Phi_D$) for Alkyl N-Arylcarbamates.
Carbamates 1a-4a have a high energy $S_2(\pi,\pi^*)\leftarrow So$ transition with a
maximum between 220 and 260 nm and a low energy $S_1(\pi,\pi^*)\leftarrow So$ transition with a maximum between 265 and 290 nm. Sample spectra of
the $S_1(\pi,\pi^*)\leftarrow So$ transition are shown for 2a and 3a in Figure 1.
Extinction coefficients for the $S_1(\pi,\pi^*)\leftarrow So$ transition of 1a-4a
are low ($\epsilon \sim 1000$) compared to the higher energy $S_2(\pi,\pi^*)\leftarrow So$ transition ($\epsilon > 10^4$). Quantum yields for decomposition ($\Phi_D$) of carbamates 1a-4a in cyclohexane are given in Table I for excitation at
254 nm and 280 nm (15). It should be noted that $\Phi_D$ for carbamates
1a-4a is essentially independent of methyl group substitution on
the phenyl ring. This is particularly surprising for 4a since
formation of both ortho and para photo-Fries products are blocked
by methyl groups substituted on the ring at the ortho and para
positions. From lifetimes for 1a ($\tau_{1a} = 3.8$ nsec) and 3a ($\tau_{3a} =$
3.9 nsec) in cyclohexane measured by Schwetlick (7) and the decomposition quantum yield in Table I, the rate constants of decomposition were calculated ($k_d = 7.1 \times 10^6$ sec$^{-1}$ for 1a and $k_d =$
$7.4 \times 10^6$ sec$^{-1}$ for 3a) and are independent of methyl substitution
on the ring.

Table I

Quantum Yield ($\Phi_D$) for Carbamate Disappearance in
Air-Saturated Cyclohexane (15)

| Carbamate | $\Phi_D$, 254 nm[a] | Concn. 254 nm | $\Phi_D$, 254 nm[a] | Concn. 280 nm |
|---|---|---|---|---|
| 1a | 0.030 | $1.15 \times 10^{-3}$ M | 0.027 (0.025)[b] | $9.44 \times 10^{-4}$ M |
| 2a | 0.022 | $1.4 \times 10^{-3}$ M | 0.024 | $6.97 \times 10^{-4}$ M |
| 3a | 0.022 | $1.08 \times 10^{-3}$ M | 0.029 (0.026)[b] | $8.07 \times 10^{-4}$ M |
| 4a | 0.026 | $1.81 \times 10^{-3}$ M | 0.024 | $1.81 \times 10^{-3}$ M |

a – All quantum yields were determined at low conversions to minimize any effects of product absorption or quenching and were
determined using GC analysis with eicosane as an internal
standard.
b – Quantum yield obtained while continuously purging with $N_2$ or
He during photolysis.

Journal of Organic Chemistry

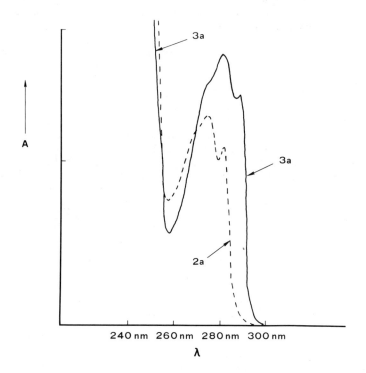

*Figure 1. UV absorption spectra (240–300-nm region) for propyl* N-p- *and* o-*tolylcarbamates (3a and 2c)*

Consideration of the effect of oxygen and excitation wavelength is important if comparisons are to be drawn between the photolysis of alkyl N-arylcarbamates under laboratory conditions and the effect of sunlight on aromatic polyurethane coatings (Table I). Oxygen has little effect on $\Phi_D$ for carbamates 1a or 3a (Table I). The results obtained at 280 nm (absorption into the $S_1(\pi,\pi^*)$ state) are of particular importance since it is this $S_1$ $(\pi,\pi^*)$ transition with maximum at 280 nm which extends above 300 nm and is responsible for absorption of ultraviolet radiation from sunlight. In addition, there is little difference in $\Phi_D$ for carbamates 1a-4a upon photolysis at 254 nm or 280 nm (Table I). These results can be interpreted from a consideration of the excited reactive states of the carbamates. The absence of a wavelength effect on $\Phi_D$ suggests an efficient internal conversion from the $S_2(\pi,\pi^*)$ state to the $S_1(\pi,\pi^*)$ state which reacts to give products. The absence of an oxygen effect suggests a short lived $S_1(\pi,\pi^*)$ state. Indeed, as noted previously, the lifetimes ($\tau_{1a}$) of 1a and 3a have been measured and are quite small ($\tau_{1a}$ = 3.8 nsec, $\tau_{3a}$ = 3.9 nsec) (7). Reaction from the triplet $\tau_1$ state has been excluded by Trecker et al. (9) who found no decrease in the product yield for 1a in the presence of triplet quenchers (e.g., cis-piperylene).

Quantum Yields ($\Phi p$) for Product Formation. The quantum yields for formation of the corresponding photoproducts b (arylamine), c (ortho photo-Fries), and d (para photo-Fries) upon photolysis of 1a-4a are given in Table II. In each case, for photolysis at 254 nm the sum ($\Phi_{total}$) of the quantum yield for photoproducts ($\Phi_b$, $\Phi_c$, and $\Phi_d$) obtained from cleavage of the N-C bond is less than the disappearance quantum yield ($\Phi_D$) calculated from the loss of starting carbamate (Table II). Since the quantum yields are quite small, any general observations relating product formation and

Table II

Quantum Yields ($\Phi p$) for Carbamate Photolysis Products -
Photolyzed at 254 nm in Cyclohexane[a] (15)

| Carbamate | ArNH$_2$ $\Phi_b$ | o-PF $\Phi_c$ | p-PF $\Phi_d$ | $\Phi_{total}$ | $\Phi_D$ |
|---|---|---|---|---|---|
| 1a | 0.005 | 0.008 | 0.003 | 0.016 | 0.030 |
| 2a | 0.005 | 0.003 | 0.008 | 0.016 | 0.022 |
| 3a | 0.002 | 0.004 | -- | 0.006 | 0.022 |
| 4a | 0.000 | -- | -- | 0.000 | 0.026 |

a - $\Phi_b$, $\Phi_c$, $\Phi_d$ are the quantum yields for formation of the aromatic amine, the ortho photo-Fries, and para photo-Fries product of the carbamate. $\Phi_{total}$ is the sum of $\Phi_b$, $\Phi_c$, $\Phi_d$. Quantum yields determined using GC analysis by comparison with known concentration of the particular product.

Journal of Organic Chemistry

methyl group substitution on the ring must be made with reservation. One observation, however, can be readily made. No 2,4,6-trimethylaniline (4b) was formed upon photolysis of 4a. Additionally, the value of $\Phi_b$ for formation of 3b is very small. Thus, substitution of the methyl group on the ring para to the carbamate group (3a and 4a) affects the reactivity of the arylaminyl radical formed upon N–C bond cleavage. This is in marked contrast to the negligible effect of methyl group substitution on the disappearance quantum yields of carbamates 2a–4a. The reaction Scheme III accounts for the results. It shows formation of the arylamine (b)

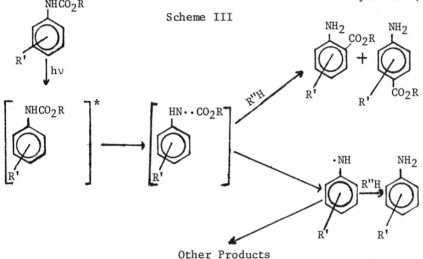

Scheme III

Other Products

and photo-Fries rearrangement products (c and d) and suggests alternate reaction paths for the arylaminyl radical – other than hydrogen abstraction. The pathway for formation of other products is particularly important for photolysis of 3a and 4a since the quantum yields for disappearance ($\Phi_D$) were comparable to the $\Phi_D$ values for 1a and 2a, despite the negligible total yields $\Phi_{total}$ for formation of photo-Fries and arylamine products. Thus, the fate of the arylaminyl radical, once formed, is significantly affected by methyl group substitution para to the nitrogen atom.

Polymer Matrix Effects. In order to approximate the environment experienced by the arylcarbamate moieties in coatings based on aromatic diisocyanates, we chose to study the photochemistry of alkyl N-arylcarbamates in polymethacrylate (PMMA) and polypropylmethacrylate (PPMA) films. First, however, 2a and 3a were irradiated in ethyl propionate (a model solvent for PMMA and PPMA) to determine the effect of the solvent polarity (dielectric) on the photolysis of the carbamates. Upon excitation at 280 nm, where the solvent absorbance was negligible, $\Phi_D$ is 0.006 for 2a and $\Phi_D$ is 0.005 for 3a. These values are significantly smaller than the values for $\Phi_D$ for 2a and 3a obtained in the non-polar cyclohexane solvent (Table I). This may be due to increases in non-radiative decay rates in the more polar ethyl propionate, a decrease in the formation rate of products from the caged radical pair induced by the ethyl propionate, or an increase in radical recombination of the arylaminyl and alkyl carboxyl radicals to give the starting carbamate. Such considerations also apply to similar results obtained by Schwetlick and co-workers (16) upon photolysis of 1a in polar solvents. Bearing in mind the results obtained in ethyl propionate, the effect of PMMA and PPMA matrices on the photochemistry of 2a is considered. A value of 0.006 for $\Phi_D$ of 2a in PMMA was obtained upon excitation at 280 nm compared to a value of 0.010 obtained in PPMA under similar conditions. These values are roughly equivalent to $\Phi_D$ for 2a in ethyl propionate ($\Phi_D$ = 0.006 at 280 nm). Variations in the methods (see experimental) used to determine absolute $\Phi_D$ values in solution and films prevents a closer comparison. However, the values of $\Phi_D$ obtained in PMMA ($\Phi_D$ = 0.006) and PPMA ($\Phi_D$ = 0.010) can be compared with confidence since the method for determining $\Phi_D$ was the same in each case. The larger value for $\Phi_D$ in PPMA versus PMMA might be explained by a consideration of the glass transition temperatures (Tg) for PPMA (Tg = 35°C) versus PMMA (Tg = 105°C). PMMA with Tg well above room temperature is a rigid matrix which can restrict the diffusion and movement of radical pairs from the cages in which they are originally formed. The flexible PPMA allows for somewhat greater movement of caged radical pairs and thus a higher $\Phi_D$ is obtained for photolysis of 2a in PPMA.

A UV analysis of the products formed upon photolysis of 2a at 280 nm in ethyl propionate, PMMA, and PPMA further illustrates the effect of the matrix stiffness on the photodecomposition process (Table III). The ratio $\Phi_c$ to $\Phi_b + \Phi_d$ [$\Phi_c/(\Phi_b+\Phi_d)$] is determined by the ratio of absorbance of product 2c to the absorbances of products 2b and 2d [$A_{2c}/(A_{2b}+A_{2d})$]. In this case, since the results were tabulated from the actual absorption spectra (difference spectra), the ratio of the products formed in the solvent ethyl propionate can be directly compared to the ratios in PPMA and PMMA. From Table III, it is readily seen that the ratio increases on going from the ethyl propionate solution,

Table III

UV Spectra Observations for Photolysis of Propyl
N-o-Tolyl Carbamate (2a)[a]

| Solvent or Matrix | $A_{2c}/(A_{2b}+A_{2d})$ |
|---|---|
| Ethyl Propionate | 1.0 |
| PPMA | 2.0 |
| PMMA | 3.0 |

a – $A_{2c}/(A_{2b}+A_{2d})$ is the ratio of the absorbance of 2c to the absorbance of 2b plus the absorbance of 2d upon photolysis of 2a.

to the flexible PPMA matrix, to the rigid PMMA matrix.  These results are reasonable since the ortho photo-Fries product 2c requires little radical mobility to form compared to 2b or 2d which
require considerable movement or diffusion to form.  In summary,
the stiffness (determined by Tg) of the solvent matrix is important in determining the distribution of products upon photolysis
of the carbamate 2a.  This must be taken into account when considering model systems for the photodegradation process of aromatic
polyurethane coatings.

   Arylamine Photodecomposition.  A number of researchers have
alluded to the fact that the products produced from photolysis of
aromatic carbamates (i.e., 1a) also degrade upon irradiation (10),
17).  Indeed, we found that the aryl amine 2b and the photo-Fries
products 2c and 2d (resulting from photolysis of 2a) decomposed
with respective disappearance quantum yields of 0.035, 0.004, and
0.003 when irradiated at 280 nm.  These latter results agree with
those of Schwetlick et al. (17), who found the rates of disappearance of 1c and 1d to be quite small.
   Due to the large quantum yield for disappearance of 2b, and
since arylamines might also be present in large quantities in
polyurethane coatings based on aromatic diisocyanates (i.e., TDI),
the disappearance quantum yields ($\Phi_D$) for the arylamines 1b-4b
were measured (Table IV).  It is obvious that methyl group substitution enhances $\Phi_D$ for the arylamines with significant increases
found for 3b and 4b which have methyl groups substituted on the
ring para to the amino group.

Table IV

Disappearance Quantum Yield for Aromatic Amines

| Aromatic Amine | $\Phi_D$, 280 nm[a] |
|---|---|
| 1b | .007 |
| 2b | .010 |
| 3b | .036 |
| 4b | .031 |

a - $\Phi_D$ values obtained in both air and helium saturated solutions
   in cyclohexane.  Concentrations for arylamine were less than
   $10^{-3}$ M in each case.

   In order to understand these results it is necessary to consider the nature of the intermediates formed upon photolysis of
arylamines.  The absorption spectra of transients produced upon
photolysis of aniline and various alkyl ring-substituted arylamines was obtained by Land and Porter (18) in different solvents
using a flash photolysis apparatus.  On this basis they identified
both an anilinyl radical (PhNH·) and an anilinyl radical cation
(PhNH$_2^+$).  The radical cation is present in polar media (H$_2$O) but
absent in cyclohexane.  From these results, a homolytic cleavage

of the NH bond was proposed as the primary process for photolysis of arylamines in non-polar media. The absence of an oxygen effect (Table IV) on $\Phi_D$ for photolysis of 1b-4b is indicative of a rapid cleavage of this NH bond and a rapid reaction of the resultant radical to give products.

Molecular Orbital Description of Arylaminyl Radicals. Arylaminyl radicals, as previously discussed, are intermediates in both the photolysis of alkyl N-arylcarbamates ($\underline{7}$, $\underline{8}$) and the photolysis of arylamines ($\underline{18}$). A simplified mechanism for photolysis of arylamines and alkyl N-arylcarbamates is illustrated in Scheme IV for the general case. An indication of the reactivity of the

Scheme IV

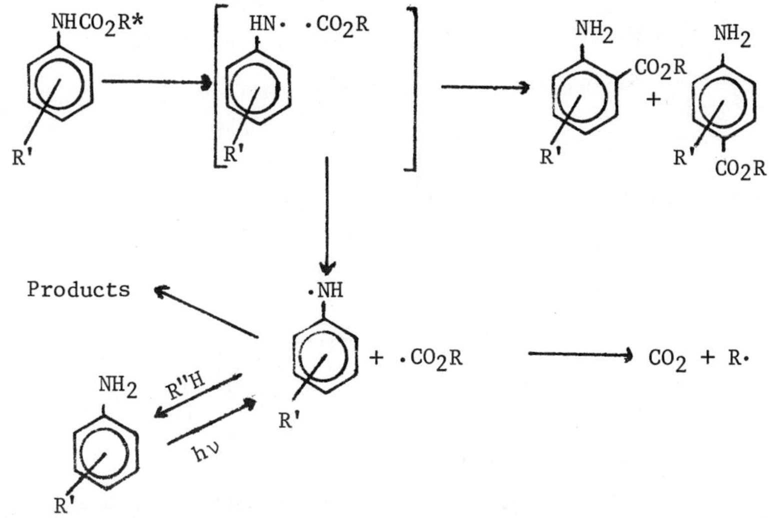

proposed common arylaminyl radical intermediate can be obtained from a molecular orbital description of the electron density of the radical. It has been reported that the reactivity orientation of radicals is largely determined by the distribution of the single electron of highest energy, the frontier electron ($\underline{19}$). An INDO molecular orbital description of the probability distribution of the frontier electron (SO MO - single occupied molecular orbital) for several arylaminyl radicals is given in Table V. In all cases, the frontier electron density is highest at the nitrogen atom, which is certainly not surprising. It is interesting, however, that methyl substitution on the ring significantly reduces the localization of this electron at the nitrogen atom. Concurrently, there is substantial electron density on the ring-substituting methyl group. This is shown in Figure 2 for the p-toluidinyl radical. It is seen that the electron density is quite large on the two out of plane hydrogen atoms attached to the ring-substituting methyl group.

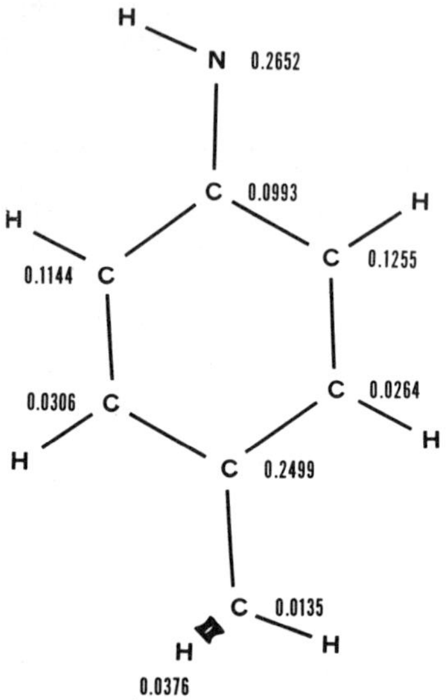

*Figure 2. Frontier electron density distribution for the p-toluidinyl radical*

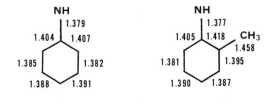

*Figure 3.   Calculated bond distance (Å) in anilinyl and methyl-substituted anilinyl radicals*

Table V

Probability of Finding the Frontier Electron at
the Various Sites in the Arylaminyl Radicals

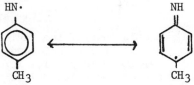

| 2 | 4 | 6 | N | C1 | C2 | C3 | C4 | C5 | C6 |
|---|---|---|---|----|----|----|----|----|----|
| H | H | H | 0.3376 | 0.0994 | 0.1437 | 0.0156 | 0.2439 | 0.0197 | 0.1400 |
| H | H | CH$_3$ | 0.2930 | 0.1076 | 0.0863 | 0.0400 | 0.2111 | 0.0008 | 0.1952 |
| H | CH$_3$ | H | 0.2652 | 0.0993 | 0.1255 | 0.0264 | 0.2499 | 0.0306 | 0.1144 |
| CH$_3$ | CH$_3$ | CH$_3$ | 0.2442 | 0.1121 | 0.1243 | 0.0207 | 0.2169 | 0.0183 | 0.1215 |

Methyl substitution on the ring, by reducing the frontier electron density at the nitrogen atom, decreases the probability for reaction at that site. Furthermore, the possibility of non-aromatic products arising from interaction at the site of ring substitution is increased. This is particularly true for the p-toluidinyl and 2,4,6-trimethylanilinyl radicals, since the probability for the frontier electron being at the ring carbon para to the site of nitrogen substitution is nearly the same as that for its' being at the nitrogen itself. If one uses the symbolism of resonance theory, the geometries and frontier electron density distributions of the p-toluidinyl and 2,4,6-trimethylanilinyl radicals indicate in each case a significant contribution from a non-aromatic canonical form. For example, the following may be written for the p-toluidinyl radical.

Although non-aromatic resonance forms can be drawn for the anilinyl and o-toluidinyl radicals, the calculated geometries and frontier electron distribution do not support such structures to the same extent as above (Figure 3). In the case of the p-toluidinyl and 2,4,6-trimethylanilinyl radicals the results suggest a decreased ability for hydrogen abstraction by the nitrogen while increasing the probability of reactions appropriate to a non-aromatic canonical form. The lowered electron density at the nitrogen for these radicals with methyl groups substituted on the ring para to the aminyl radical accounts for the low quantum yields for formation of 3b (0.002) and 4b (0.000) upon photolysis

of 3a and 4a.  Additionally, the lowered electron density at the
nitrogen atom of the p-toluidinyl and 2,4,6-trimethyl anilinyl
radicals accounts for the higher quantum yields for disappearance
of 3b and 4b (Table III).  Once formed, the p-toluidinyl and 2,4,6-
trimethyl anilinyl radicals will react to give products in lieu of
hydrogen abstraction and return to starting arylamine (3b or 4b).
Arylaminyl radicals formed from photolysis of 1b and 2b with no
para methyl groups substituted on the phenyl ring, have the electron density predominantly on the nitrogen atom and tend to abstract a hydrogen and return to the starting arylamine (1b or 2b).
Thus the quantum yields for decomposition for 1b and 2b are quite
low (Table III).  In summary, the electron distribution of the
intermediate arylaminyl radical accounts for both the negligible
quantum yield for formation of the amines 3b and 4b upon photolysis of carbamates 3a and 4a as well as the higher quantum yields
for disappearance of amines 3b and 4b compared to 1b and 2b.

   Up to this point we have discussed only carbamates 1a-4a with
a single carbamate group on the phenyl ring as model systems for
aromatic polyurethane photodecomposition.  In polyurethane coatings based on the aromatic diisocyanate TDI two carbamate groups
are attached to the phenyl ring.  Furthermore commercially available TDI is actually a mixture of 2,4-toluene diisocyanate (2,4-
TDI) and 2,6-toluene diisocyanate (2,6-TDI) which when formulated
give 2,4- and 2,6-biscarbamates.  Model systems for these species
would then be biscarbamates of 2,4-TDI and 2,6-TDI (as shown
below) and not carbamates such as 1a-4a.

Biscarbamate of 2,4-TDI          Biscarbamate of 2,6-TDI

   Molecular orbital calculations of the radicals produced by
photochemical cleavage of the various N-C bonds of the biscarbamates of 2,4-TDI and 2,6-TDI should provide information about the
reactivity of these resultant radicals.  In this way, differences
between the carbamates 1a-4a and biscarbamates can be predicted.
The frontier electron distributions of the two possible arylaminyl
radicals formed by photolytic N-C bond cleavage in the biscarbamate of 2,4-TDI (R=CH$_3$) are shown in Figures 4 and 5.  In both
radicals, the electron density is largest at the ring carbon of
methyl group substitution.  Furthermore, densities on the nitrogen
atom at the point of N-C bond cleavage are quite small (.17 in
both cases) compared to the large electron densities on the nitrogen atom of the anilinyl and even the p-toluidinyl radicals (Table
V) produced by N-C bond cleavage of the carbamates 1a and 3a respectively.  These results suggest that products would arise from

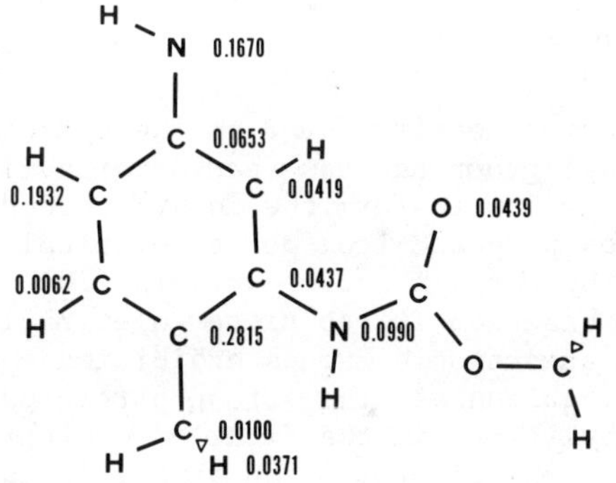

*Figure 4. Frontier electron density distribution for the 3-methylcarbamyl-4-methyl anilinyl radical*

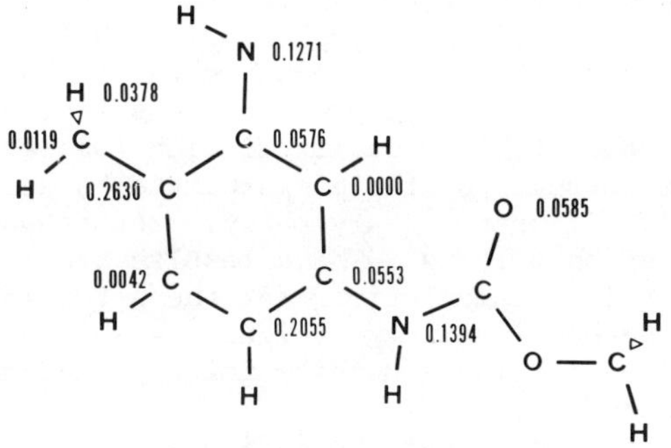

*Figure 5. Frontier electron density distribution for the 2-methyl-5-methylcarbamyl anilinyl radical*

these intermediate radicals not from hydrogen abstraction by the
nitrogen atom, but by elimination of a hydrogen atom from the ring
substituted methyl group to give products of quirone-like structure. Specifically, it can be concluded that this follows from

the frontier electron density found on the hydrogen atoms of the
substituting methyl group and the ready availability of the electron on the ring carbon to form the double-bonded $CH_2$ group. The
quinoid structures proposed from our theoretical studies support
earlier findings by Nevskii and co-workers (20) who on exposure
of polyurethanes based on TDI to ultraviolet radiation identified
the formation of auxochromic groups exhibiting chemical properties
characteristic of quinones. The strong participation of the ring
substituting methyl group in the frontier orbital system of the
arylaminyl radicals in Figures 4 and 5 is critical to the formation of quinoid-like products. Thus, it is not surprising that
Schollenberger found the polyurethane based on m-phenylene diisocyanate (PDI) with no ring substituted methyl groups to be
color stable compared to TDI based polyurethanes (3).

Results that we obtained can also be explained by the molecular orbital calculations. Photolysis of either carbamate 3a or
the bispropylcarbamate of 2,4-TDI (5a) resulted in a loss of the
integrated NMR signal of the phenyl protons ($H_x$), the amino
protons ($H_a$), and the arylmethyl protons ($H_e$) compared to the
integrated areas of the propyl group protons ($H_b$, $H_c$, $H_d$), which
appeared relatively stable upon irradiation (Tables VI and VII).
This effect was largest for the biscarbamate 5a as no NMR signal
from the phenyl, nitrogen, or arylmethyl protons was obtained
after 78 hours of photolysis. These results are consistent with
the molecular orbital calculations for the p-toluidinyl (Table V)
and the methyl carbamyl anilinyl (Figures 4 and 5) radicals and
suggest formation of quinone methine imide photodegradation
products for 3a and 5a.

It might be expected that the radicals produced by N-C bond
cleavage of a biscarbamate based on 2,6-TDI would be different
from similar radicals obtained from a 2,4-TDI based biscarbamate.
From Figure 6, observations can be readily made concerning the
radical produced by N-C bond cleavage of either of the carbamate
groups of the biscarbamate of 2,6-TDI. The substituting methyl
group is now connected to a ring carbon atom with no appreciable
electron density. Additionally, the methyl group itself, and the
out-of-plane hydrogen atoms on the methyl group, have negligible
frontier electron density. Thus, it is highly improbable that
quinone-like products would result from this radical. This may

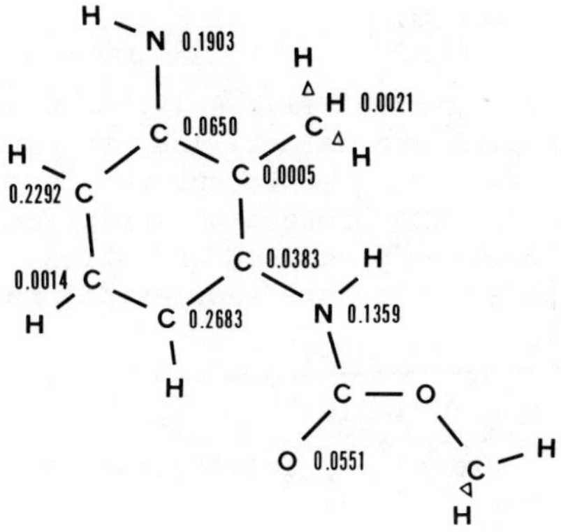

*Figure 6. Frontier electron density distribution for the 2-methyl-3-methylcarbamyl anilinyl radical*

account for the fact that polyurethanes based on TDI with a high content of 2,4-TDI discolor to a greater extent than coatings composed primarily of the 2,6-TDI isomer ($\underline{20}$).

Table VI

Photolysis of the Bispropyl Carbamate of 2,4-TDI (5a)[a,b]

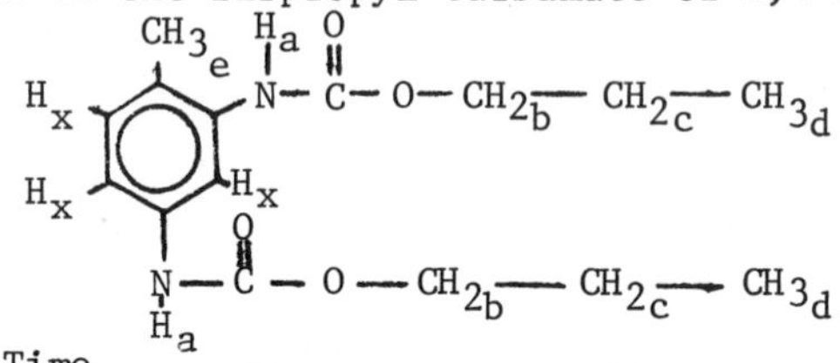

$$H_x \underset{H_x}{\overset{CH_3}{\bigcirc}} \overset{H_a}{\underset{e}{N}} - \overset{O}{\overset{\|}{C}} - O - CH_{2b} - CH_{2c} - CH_{3d}$$

$$\underset{H_a}{N} - \overset{O}{\overset{\|}{C}} - O - CH_{2b} - CH_{2c} - CH_{3d}$$

| Photolysis Time (Hours) | $H_a/(H_b+H_c+H_d)$ | $H_x/(H_b+H_c+H_d)$ | $H_e/(H_b+H_c+H_d)$ |
|---|---|---|---|
| 0 | .13(.14)[c] | .20(.21)[c] | .20(.21)[c] |
| 23 | .09 | .15 | .15 |
| 27 | .07 | .13 | .14 |
| 78 | .00 | .00 | .00 |

a - A .05 M undegassed acetonitrile solution was irradiated with a 200 Watt medium pressure mercury lamp through a pyrex filter.

b - $H_a$, $H_x$, $H_e$, $H_b$, $H_c$ and $H_d$ represent the integrated areas of the NMR signals for the protons as indicated in the structural diagrams. The data are presented as ratios.

c - The values in parenthesis are theoretical ratios.

Table VII

Photolysis of Propyl N-p-Tolylcarbamate (3a)[a,b]

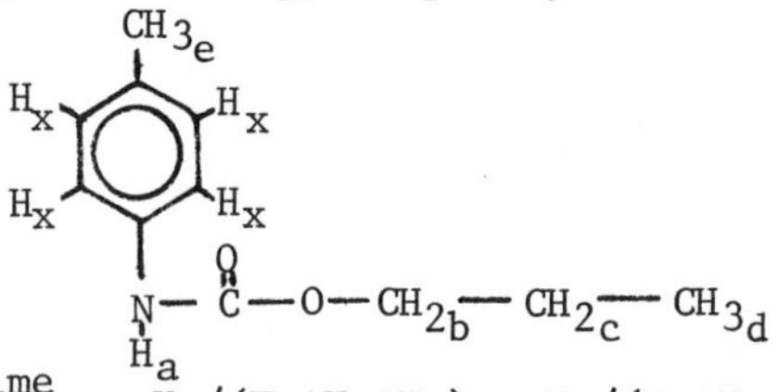

$$H_x \underset{H_x}{\overset{CH_{3e}}{\bigcirc}} H_x$$

$$\underset{H_a}{N} - \overset{O}{\overset{\|}{C}} - O - CH_{2b} - CH_{2c} - CH_{3d}$$

| Photolysis Time (Hours) | $H_a/(H_b+H_c+H_d)$ | $H_x/(H_e+H_c+H_d)$ | $H_e/(H_b+H_c+H_d)$ |
|---|---|---|---|
| 0 | .14(.14)[c] | .61(.57)[c] | .41(.43)[c] |
| 66 | .09 | .36 | .26 |

a - A .1 M undegassed acetonitrile solution was irradiated with a 200 Watt medium pressure mercury lamp through a pyrex filter.

b - $H_a$, $H_x$, $H_e$, $H_b$, $H_c$, and $H_d$ represent the integrated areas of the NMR signals for the protons as indicated in the structural diagrams. The data are presented as ratios.

c - The values in parenthesis are theoretical ratios.

In conclusion, it has been shown that methyl groups substituted on the phenyl ring of alkyl N-arylcarbamates and arylamines affect the photochemistry. The photolysis of alkyl N-arylcarbamates was found to be dependent on the rigidity of the solvent matrix in which the carbamate was dissolved. The reactivity of arylaminyl radicals, intermediate in both the photolysis of arylamines and alkyl N-arylcarbamates, was described by the molecular orbital frontier electron density distribution. Molecular orbital calculations made on biscarbamates of 2,4-TDI and 2,6-TDI were used to explain a number of experimental observations dealing with the photolysis of polyurethane coatings based on TDI isomer mixtures. Finally, it should be made quite clear that the results obtained from the photolysis of carbamates 1a-4a must be viewed with reservation when these carbamates are used as model systems for the photodecomposition of polyurethane coatings based on aromatic diisocyanates.

## Literature Cited

1.  Schollenberger, C. S.; Dinbergs, K.; S.P.E. Transactions, 1961, 1, 31.
2.  Nevskii, L. V.; Tarakanov, O. G.; Beljakov, O. K., Soviet Plastics, 1966, 7, 45.
3.  Schollenberger, C. S.; Stewart, F. D., Adv. Urethane Sci. and Technol., 1973, 2, 71.
4.  Allen, N. S.; McKellar, J. F., J. Appl. Polym. Sci., 1976, 20, 1441.
5.  Schollenberger, C. S.; Stewart, F. D., Adv. Urethane Sci. and Technol., 1975, 4, 66.
6.  Bellus, D.; Schaeffer, K., Helv. Chim Acta, 1968, 51, 221.
7.  Schwetlick, K.; Noack, R., Tetrahedron, 1979, 35, 63.
8.  Schwetlick, K.; Noack, R.; Schmieder, G., Z. Chem., 1972, 12, 107.
9.  Trecker, D. J.; Foote, R. S.; Osborn, C. L., Chem. Commun., 1968, 1034.
10. Beachell, H. C.; Chang, I., J. Polymer Sci., Al, 1972, 10, 503.
11. Masilamani, D.; Hutchins, R. O.; Ohr, J., J. Org. Chem., 1976, 41, 3687.
12. Turro, N. J., in "Molecular Photochemistry", W. A. Benjamin, Inc., Reading, Mass., 1965, 6.
13. Pople, J. A.; Beveridge, D. L. in "Approximate Molecular Orbital Theory", McGraw-Hill, New York, 1970, Chapter 4.
14. Purcell, K. F.; Zapata, J., QCPE, 1976, 11, 312.
15. Herweh, J. E.; Hoyle, C. E., J. Org. Chem., 1980, 11, 2195.
16. Schwetlick, K.; Noack, R., Z. Chem., 1972, 12, 143.
17. Schwetlick, K.; Noack, R., Z. Chem., 1972, 12, 140.
18. Land, E. J.; Porter, G., Trans. Faraday. Soc., 1963, 59, 2027.
19. Fukui, K., "Theory of Orientation and Stereoselection", Springer-Verlag, New York, 1965, Chapter 7.
20. Nevskii, L. V.; Tarakanov, O. G., Soviet Plastics, 1967, 9, 47.

RECEIVED September 16, 1980.

# Photodegradation and Photoconductivity of Poly(N-vinylcarbazole)

ROBERT F. COZZENS

George Mason University, Fairfax, VA 22030

The photochemistry of macromolecules is often dominated by the photophysical processes that occur after initial photon absorption and prior to the chemical reactions leading to actual molecular degradation. The transport of energy from one location within a material to some distant site where chemical or physical interactions may occur is of considerable interest in the fields of electrical photoconductivity, solar energy conversion, UV-curing of polymers and polymer photodegradation. Energy transfer in polymeric materials can occur by both an intramolecular and intermolecular mechanism and may involve both singlet and triplet species. Without energy transfer, interaction with the excited state could occur only at the site of initial photon absorption. Energy transfer allows chemical species other than those initially absorbing a photon to be responsible for the processes of interest.

## Photoconductivity of Polymers

In order for a material to conduct electricity it is necessary that charge carriers (positive and/or negative) be present and able to migrate under the influence of a polarizing electric field. The transport of charge, just as the transport of energy, need not involve the displacement of mass but may be viewed as a wave phenomenon. For a polymer to be photoconductive the absorption of a photon must lead to the formation of a mobile charge carrier. Energy transfer processes may precede the generation of carriers. Charge carrier migration involves the transfer of an electron between neighbors, and is analogous to resonance

0097-6156/81/0151-0137$05.00/0

transfer of energy.  Carrier migration may be viewed
as either the movement of a hole or positive charge
toward a negative electrode or the transport of an
electron toward a positive electrode.  Charge transport is described in Figure 1.  Considerable interest
has been focused on photoconductive polymers by the
electrophotography industry since the first report (1)
on these materials.  The list of sensitizing additives
has been continually expanding since those first
studies (2).  The subject of polymer photoconduction
has been extensively reviewed in recent years (3).

     The most extensively studied photoconductive
polymer is poly(N-vinylcarbazole) PVCa.  The accepted
mechanisms for photogeneration of charge carriers in
PVCa films involves localization of migrating excitation energy at a trapping site, followed by electron
transfer to a neighboring group.  The resulting charged
geminate pair may then separate and either the hole or
electron or both may migrate in a polarizing field.  If
both the hole and electron are mobile, as appears to be
the case in undoped, purified PVCa, the possibility of
collision and subsequent geminate pair recombination is
increased.  Doping of the polymer with appropriate dyes
or charge-transfer complex forming agents increases the
efficiency of carrier generation, reduces the probability of recombination and extends the effective range
of photoresponse into the visible region of the spectrum.  The most commonly used electron acceptor com-
plexing agent is 2,4,7-trinitro-9-fluroenone (TNF) (4),
although many other electron accepting groups have been
studied.  Measurements of charge mobility in PVCa-TNF
films show that charge carriers are predominately
electrons (4b).  Mechanisms of dye sensitization (5)
involve both energy transfer from the dye to the polymer and/or electron transfer from a trapped exciton to
a dye molecule leading to the formation of a charge
carrier.

## Degradation of Poly(N-vinylcarbazole)

     In the case of undoped PVCa films, impurities and
surface states dominate the photoconduction mechanism
(6) leading one to question any study of intrinsic
photoconduction in organic polymers.  Poly(N-vinylcarbazole) films yellow under ambient laboratory conditions.  Work in our laboratory (7) has shown that
ageing of a purified sample of PVCa leads to an increase in photoresponse in the 350-450 nm region while
there is an initial  drop in photoresponse in the 250-

300 nm region. This is demonstrated in Figure 2. Recent work (8) has shown that exposure of PVCa films to
ultra-violet radiation results in the formation of carbonyl groups at or near the surface. The carbonyl
groups act as singlet exciton traps and compete with
excimer formation for migrating energy. It is proposed
(8) that these photo-oxidation products form exciplexes
with excited carbazole chromophores and probably serve
as electron-accepting traps. Small quantities of these
photo-oxidation products have a marked effect on the
photoconductivity of otherwise "pure" PVCa films that
leads to an increase in the photocurrent in the 300-
450 nm spectral region by a factor of 15 (7,8).

The photodegradation of PVCa solutions has been
studied to determine a mechanism for the photodegradation process and identify the excited states involved.
The polymer was secondary grade (Aldrich Chemical
Company) that was further purified by reprecipitation
three times from methylene chloride solution by the
dropwise addition into methly alcohol. Solutions in
methylene chloride were found to yellow rapidly in air
upon exposure to 360 nm light, whereas solutions degassed by the freeze-pump-thaw technique showed little
yellowing.

Monochromatic radiation at 344 nm, 294 nm and 261
nm was provided by a monochromater coupled with a 150
watt xenon arc. The selected wavelengths of radiation
correspond to absorption by the three lowest singlet
states of the carbazole chromophore. Sample solutions
were irradiated in equilibrium with air. Quantitative
measurements of yellowing were made by observing increases in absorption at 390 nm. The standard ferrioxalate technique was employed to determine the relative
quantum efficiency of yellowing at each of these wavelengths. All three singlet states had the same quantum yield for increase in absorption at 390 nm
following irradiation, leading one to consider the lowest singlet or triplet state produced by the rapid
internal conversion from the higher singlet states, as
responsible for yellowing. To determine which of these
two states are responsible for photo-oxidative
yellowing, triplet quenchers were added and Stern-
Volmer plots were obtained. Both piperylene and
naphthalene were effective in substantially reducing
the rate of yellowing when samples were monochromati-
cally irradiated at 344 nm. Stern-Volmer plots are
shown in Figure 3.

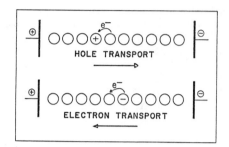

Figure 1. Diagram of charge carrier transport in conductive materials

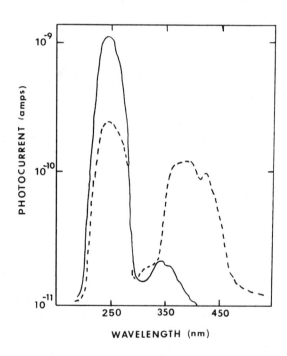

Figure 2. Photoconductive response at 5,000 V/cm of fresh (————) and photo-degraded (– – –) poly(N-vinylcarbazole) films as a function of wavelength of excitation

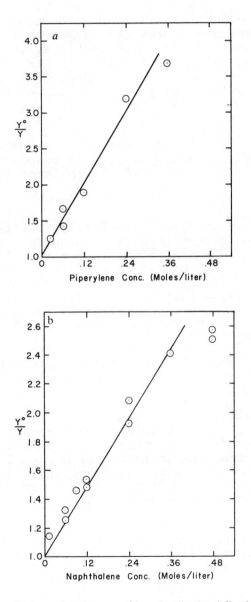

*Figure 3.   Stern–Volmer plots for quenching of yellowing following photolysis of PVCa solutions in methylene chloride by (a) piperylene and (b) naphthalene.   Yellowing is measured as the increase in absorption at 390 nm.*

Phosphorescence of solutions of freshly prepared PVCa in methyltetrahydrofuran in the glassy state at 77°K was observed both with and without piperylene or naphthalene present. The relative energies of the singlet and triplet states of naphathalene and piperylene compared to those of carbazole make energy trans fer from the triplet state of carbazole to either of these two compounds energetically allowed while transfet from the carbazole singlet is energetically prohibited. Both compounds were found to effectively quench phosphorescence confirming the effectiveness of these materials as triplet quenchers. It was concluded that the lowest triplet state of PVCa is responsible for photo-oxidative yellowing of this polymer in solution.

Delayed emission spectra from fresh and photodegraded solutions of PVCa in methyltetrahydrofuran at 77 °K are shown in Figure 4. Freshly purified material shows phosphorescence bands typical of the carbazole chromophore and delayed fluorescence resulting from annihilation of two migrating triplets. Such spectra are typical of aromatic polymers where energy migration is efficient (9). Photodegraded PVCa solutions exhibit only a broad band in the delayed emission spectrum indicating efficient trapping of migrating energy by degradation products.

Infrared absorption spectra of degraded PVCa show distinct carbonyl bands; these bands are not present in the starting material. Some crosslinking (ca. 4% by weight of the original polymer) was detected as insoluble residue. There was an average decrease in molecular weight of the soluble polymer following photodegradation of less than 10%, measured viscometrically. The identification of carbonyl products is consistent with the proposed mechanism of degradation of PVCa films by Pfister and Williams (6) and Itaya, Okamoto and Kusabayashi (8).

Films of purified PVCa were cast from methylene chloride solution on quartz plates. The solvent was allowed to slowly evaporate to give smooth, clear films with a thickness of ca. 5.0 nm. Contact angle measurements using water droplets were measured with a standard contact angle goniometer. Samples were photolysed in air with polychromatic light from a 150 watt xenon arc. Contact angles were measured after various times of irradiation to monitor the formation of oxidation products at the surface of the polymer films.

Results are shown in Figure 5. Samples of PVCa were doped with 10% by weight of poly(1-vinylnaphthalene) to determine if the naphthalene chromophore would serve as a quencher for the surface oxidation of PVCa as it appears to do in the case of fluid solutions. The data in Figure 5 indicates that the presence of the naphthalene chromophore does not prevent the decrease in contact angle observed upon the irradiation of PVCa. Thus, naphathalene may not be an effective quencher of the surface oxidation of PVCa film as it appears to be for the yellowing of PVCa in solution. It is possible that such an effect may be masked by the decrease in contact angle upon the photolysis of poly(2-vinylnaphthalene) films as shown in Figure 5. The surface oxidation products detected by the decrease in contact angle upon photolysis of PVCa films may dominate the photoconductivity of this polymer. Work is underway to confirm this relationship and measure surface conductivity simultaneously with bulk conductivity as a function of photodegradation.

Conclusions

It may be concluded that the solution photolysis of PVCa results in the formation of oxidation products assumably responsible for yellowing. The mechanism of photo-oxidation involves the triplet state of the polymer and quenching of this state reduces the observed rate of yellowing. In the case of solid films of PVCa, photolysis leads to the formation of polar groups at the surface of the polymer film resulting in a decrease in the contact angle of water with the film surface. The triplet quenching chromophore naphthalene does not seem to effectively quench the formation of these surface states indicating that the mechanism of photo-oxidation of the solid phase of PVCa may be different than that in solution.

Work is continuing to correlate the formation of surface oxidation states with changes in the photoconduction of films of PVCa. The relationship between energy transfer and photoconductivity is being investigated.

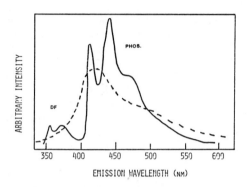

*Figure 4.   Delayed emission of purified (———) and photodegraded (— — —) poly(N-vinylcarbazole) solutions in methyltetrahydrofuran at 77 K when excited at 290 nm*

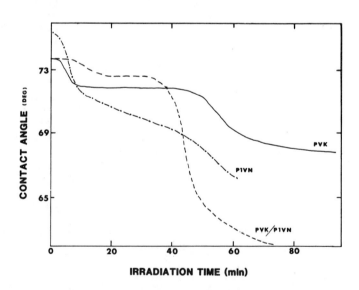

*Figure 5.   Contact angle of water with films of poly(N-vinylcarbazole) (———), poly(1-vinylnaphthalene) (— · — ·), and a 10 wt % mixture of the two (— — —) as a function of time of photolysis in air*

## References

1. H. Hoegl, et. al., U.S. Patent 3037861, Sept. 8 (1958).

2. a) H. Hoegl, J. Phys. Chem., 69, 755 (1965).

   b) M. Landon, E. Lell-Doller, and J. Weigl, Molecular Crystals, 3, 241 (1967).

3. a) F. Gutmann and L.E. Lyons, "Organic Semiconduction", Wiley, New York (1967).

   b) J. M. Pearson, Pure and Applied Chemistry, 49, 463 (1977).

   c) J. Mort and D.M. Pai (Eds), "Photoconductivity and Related Phenomena", Elsevier, New York (1976).

   d) A.V. Patsis and D.A. Seanor, "Photoconductivity in Polymers: An Interdisciplinary Approach", Technomic Pub. Co., Westport, Ct. (1976).

   e) M. Hafano and K. Tanikawa, Progress in Organic Coatings, 6, 65 (1978).

4. a) H. Hoegl and Neugebauer, U.S. Patent 3162532, June 13 (1960).

   b) R.M. Schaffert, IBM J. Res. Dev., 15, 75 (1971).

   c) M.D. Shattuck and U. Vahtra, U.S. Patent 3484237, June 13 (1966).

   d) R.C. Hughes, Appl. Phys. Lett., 21, 196 (1972).

5. P.J. Reucroft, Polymer-Plast. Technol. Eng., 5, 199 (1975).

6. G. Pfister, and D.J. Williams, J. Chem. Phys., 61, 2416 (1974).

7. R.F. Cozzens and W. Mindak, 57th Meeting Virginia Academy of Sciences, Richmond, Virginia (1979) and additional results to be published.

8. A. Itaya, K. Okamoto, and S.Kusabayashi, Bull. Chem. Soc. Japan, 52, 2218 (1979).

9. a) R. F. Cozzens, and R.B. Fox, J. Chem. Phys., 50, 1532 (1969).

   b) R.B. Fox, and R.F. Cozzens, Macromolecules, 2, 181 (1969).

RECEIVED November 17, 1980.

# Effect of Metal Salts on the Photoactivity of Titanium Dioxide

## Stabilization and Sensitization Processes

GETHER IRICK, JR., G. C. NEWLAND, and R. H. S. WANG

Research Laboratories, Tennessee Eastman Company,
Division of Eastman Kodak Company, Kingsport, TE 37662

Titanium dioxide has been used for many years either alone or in combination with other pigments and fillers for the opacification of polymeric materials (1). An undesirable side-effect of titanium dioxide use is the photooxidation of oxidizable polymers in which it is incorporated (2,3); it has also been observed to decrease the lightfastness of dyes (4). Pigment manufacturers have reduced the photoactivity of titanium dioxide by minimizing metal impurities and by using various surface treatments. Crystal structure is also recognized as an important determinant of photoactivity; anatase pigments are generally more photoactive than rutile pigments (3,6). Photoreactions involving titanium dioxide are not always undesirable. This pigment has been used as a heterogeneous photocatalyst for reduction of nitrogen (7,8), for decomposition of carboxylic acids (9), and for radical polymerization (10).

All of these uses are based on the behavior of titanium dioxide as a semiconductor. Photons having energies greater than ∿ 3.2 eV (wavelengths shorter than 400 nm) produce electron/hole separation and initiate the photoreactions. Electron spin resonance (esr) studies have demonstrated electron capture by adsorbed oxygen to produce the superoxide radical ion (Scheme 1) (11). Superoxide and the positive hole are key factors in photoreactions involving titanium dioxide; reported here are the results of attempts to alter the course of these photoreactions by use of metal ions and to understand better the mechanisms of these photoreactions.

## Experimental

Chemicals. Most of the pigments and chemicals were obtained from commercial sources and were not treated before use. Exceptions were isopropyl alcohol and tert-butyl alcohol, which distilled to remove detectable carbonyl impurities.

Determination of Pigment Photoactivity by Isopropyl Alcohol Oxidation. A previously reported method (5) was used with minor

Scheme 1

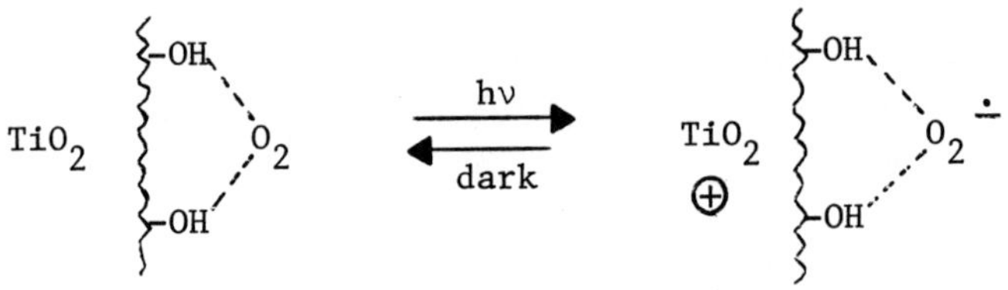

Table 1

Photoactivities of Selected Commercial Titanium Dioxide Pigments

| Pigment | Primary Crystalline Form | $\Phi_{rel}$ in Isopropyl Alcohol |
|---|---|---|
| Baker Reagent[1] | Anatase | 1.0 |
| "Titanox" AMP[2] | " | 0.6 |
| "Unitane" 0520[3] | " | 0.6 |
| "Titanox" AA[2] | " | 0.3 |
| "Ti-Pure" 33[4] | " | 0.2 |
| "Titanox" A-168-LO[2] | " | 0.08 |
| "Ti-Pure" R-110[4] | Rutile | 0.1 |
| "Ti-Pure" R-100[4] | " | 0.08 |
| "Ti-Pure" R-992[4] | " | 0.07 |
| "Titanox" RA-NC[2] | " | 0.05 |
| "Ti-Pure" R-610[4] | " | 0.02 |
| "Titanox" RA-50[2] | " | 0.008 |

1.  J. T. Baker Company
2.  Titanium Pigment Corporation
3.  American Cyanamid Company
4.  E.I. DuPont de Nemours

modifications. A slurry of 2.0 g of titanium dioxide in 8 mL of freshly distilled isopropyl alcohol (acetone-free) was placed in a 16- x 150-mm(od) culture tube of Pyrex glass, and the vapor space above the liquid was purged with oxygen. The tube was closed with a screw cap lined in Teflon fluorocarbon and rotated with inversion during irradiation by a 280-750 nm mercury lamp (General Electric Model H1000RDXFL36-15). Photon flux in the sample tube was $\sim$ 3.4 x $10^{-6}$ einsteins min$^{-1}$ (based on reagent anatase as the actinometer). Acetone analyses are reported (Table 1) relative to those obtained with reagent anatase (J. T. Baker, absolute quantum yield 0.5 at 366 nm photon flux of 2.9 x $10^{-5}$ einsteins min$^{-1}$).

Preparation of Coated Pigments. A slurry of 50 g of titanium dioxide in 100 mL of distilled water with sufficient metal salt to provide the desired coating concentration was stirred on a steam bath to constant weight. The pigment was then ground in a mortar and dried in air at 100°C for 96 hr.

Metal Analyses of Commercial Pigments. Quantitative emission spectrographic analyses were performed for 38 metals (Stewart Laboratories, Inc.). Detection limits for the 24 undetected metals are given in Table 3.

Determination of Pigment Photoactivity by Leuco Dye Oxidation. A slurry of 0.5 g of coated or uncoated titanium dioxide in 5 mL of benzene with 0.0049 mmol of the leuco dye tris[4-(N,N-dimethyl-amino)phenyl]methane (leuco crystal violet I) was oxygen-purged and irradiated for 20 min as described above for the isopropyl alcohol oxidation. The slurry was centrifuged, and the benzene layer was decanted and discarded. The pigment was extracted twice with 5 mL of methanol by shaking, centrifuging, and decanting. The extracts (10 mL) were mixed and absorbance was determined at 590 nm by using a Coleman-Hitachi Model 124 spectrophotometer (Tables 6 and 7). The use of benzene solutions of leuco crystal violet as an atomic oxygen detector was previously reported (12).

Preparation and Degradation of Polypropylene Formulations. A mixture of unstabilized polypropylene powder (Tennessee Eastman Company, I.V. 1.85 in tetrahydronaphthalene at 145°C) and pigment (coated and uncoated) was dry-blended by tumbling for at least 1 hr. The mixture was then melt-compounded in a Brabender Plasti-Corder high-viscosity recording tester (C. E. Brabender Corp.) and granulated. Films were pressed, and 0.5- x 2.5- x 0.005-in. strips were mounted on white cardboard. The films were exposed in a Uvatest Model 110 weathering device (Geopar Industries) equipped with a bank of 40-W fluorescent lamps with emissions centered at 310 and 366 nm. Specimens were evaluated for brittle-ness by bending them 180° around a 4-mm mandrel with the exposed

side of the film facing outward.  Samples were run in duplicate,
and the failure times reported are the average for two breaks
per specimen (Table 8).

Determination of Additive Effects on the Decomposition of
tert-Butyl Hydroperoxide and Hydrogen Peroxide.  Solutions of
tert-butyl hydroperoxide (1.0 mmol) and 30% aqueous hydrogen
peroxide (1.32 mmol) in 5 mL of tert-butyl alcohol with the
various additives (Tables 9 and 10) were held at 80°C for 24 hr.
Peroxide analyses were obtained by sodium iodide/0.05N sodium
thiosulfate titration.

Discussion

Photoactivities of Commercial Pigments.  The photoactivities
of titanium dioxide pigments, as indicated by the quantum yields
for photooxidation of isopropyl alcohol to acetone, vary by a
factor of $10^3$ (Table 1).  The observation of generally lower
photoactivity of the rutile relative to the anatase pigment is
consistent with previous observations (5).

Effect of Metal Salts on Reagent Anatase Photoactivity.
The most active of these pigments, a reagent-grade anatase, was
used for screening a group of metal acetates for possible deactivation effects.  Acetates were used because of their solubility
and because the acetic acid released upon reaction with the
pigment surface could be removed by heating.  Of the 31 metals
investigated, four (cerium, zinc, cobalt, and manganese) func-
tioned as very efficient deactivators (Table 2).  In fact, the
activity of the treated anatase was equal to or, in some cases,
lower than that of all the commercial anatase and most of the
commercial rutile pigments.

Metal Analyses of Commercial Pigments.  While the surface
coating of pigments with alumina-silica to control photoactivity
and to provide other desirable properties is well known (13), it
was interesting to speculate whether trace metals were also being
used commercially for this purpose.  Incorporation of cobalt (14)
and manganese into polyesters and manganese into nylon (15) has
been reported to impart increased weatherability.  Analyses of the
group of commercial pigments for 38 metals, including the four
found to decrease pigment photoactivity, were not totally conclusive but suggested that zinc was used as a deactivator
(Table 3). Cerium and cobalt were not detected (lower detection
limits of 0.01 and 0.004 wt %, respectively); manganese was
present at a concentration of 0.001% or less, and zinc was present
at significant concentrations (> 0.1%) in three of the four least
active pigments evaluated.  Of the three pigments containing zinc,
two also had high levels of aluminum, indicative of a surface
alumina coating.  Thus only one pigment in the group exhibited
a low photoactivity that could be accounted for solely on the
basis of its zinc content (2.5 wt %).

Table 2

Photoactivity of Reagent Anatase Titanium Dioxide
Coated with Metal Acetates

| Metal Acetate Coating, 3 wt % | $\Phi_{rel}$ In Isopropyl Alcohol |
|---|---|
| None | 1.0 |
| Silver | 0.9 |
| Thallium | 0.9 |
| Gallium | 0.9 |
| Ferric | 0.6 |
| Lead | 0.6 |
| Rubidium | 0.5 |
| Strontium | 0.5 |
| Aluminum | 0.5 |
| Lanthanum | 0.5 |
| Zirconium | 0.4 |
| Uranyl | 0.4 |
| Potassium | 0.4 |
| Samarium | 0.4 |
| Praseodymium | 0.3 |
| Niobium | 0.3 |
| Neodymium | 0.3 |
| Cupric | 0.3 |
| Magnesium | 0.3 |
| Barium | 0.3 |
| Yttrium | 0.3 |
| Sodium | 0.3 |
| Lithium | 0.2 |
| Chromic | 0.2 |
| Stannous | 0.2 |
| Didymium | 0.2 |
| Nickelous | 0.2 |
| Calcium | 0.2 |
| Cerous | 0.02 |
| Zinc | 0.01 |
| Cobaltous | 0.01 |
| Manganous | < 0.001 |

Table 3

Metal Analysis of Commercial Titanium Dioxide Pigments[a]

| Titanium Dioxide | $\Phi_{rel}$ | Al | Ca | Cr | Cu | Fe |
|---|---|---|---|---|---|---|
| Baker Reagent | 1.0 | .01 | .004 | <.0004 | .0002 | .007 |
| "Titanox" AMP | .6 | 3.0 | .004 | <.0004 | .0004 | .004 |
| "Unitane" 0520 | .6 | 1.3 | .01 | <.0004 | .0004 | .007 |
| "Titanox" AA | .3 | .1 | .004 | <.0004 | .0002 | .010 |
| "Ti-Pure" 33 | .2 | .1 | .004 | <.0004 | .0002 | .004 |
| "Titanox" A-168-LO | .08 | .25 | .004 | <.0004 | .0002 | .004 |
| "Ti-Pure" R-110 | .1 | .1 | .004 | <.0004 | .0007 | .010 |
| "Ti-Pure" R-100 | .08 | .1 | .004 | <.0004 | .0004 | .004 |
| "Ti-Pure" R-992 | .07 | .1 | .004 | <.0004 | .0004 | .005 |
| "Titanox" RA-NC | .05 | 1.3 | .004 | <.0004 | .0002 | .004 |
| "Ti-Pure" R-610 | .02 | .09 | .007 | <.0004 | .0004 | .004 |
| "Titanox" RA-50 | .008 | 2.5 | .004 | <.0004 | .0004 | .004 |

| Titanium Dioxide | $\Phi_{rel}$ | Mg | Mn | Nb | Ni | Pb |
|---|---|---|---|---|---|---|
| Baker Reagent | 1.0 | .0025 | <.0004 | .004 | <.001 | .001 |
| "Titanox" AMP | .6 | .0025 | .0007 | .004 | <.001 | .001 |
| "Unitane" 0520 | .6 | .005 | <.0004 | .004 | <.001 | .007 |
| "Titanox" AA | .3 | .004 | <.0004 | .004 | <.001 | <.001 |
| "Ti-Pure" 33 | .2 | .0025 | <.0004 | .004 | <.001 | .004 |
| "Titanox" A-168-LO | .08 | .001 | <.0004 | .004 | <.001 | .001 |
| "Ti-Pure" R-110 | .1 | .010 | <.0004 | .02 | <.001 | .004 |
| "Ti-Pure" R-100 | .08 | .001 | <.0004 | .004 | <.001 | .001 |
| "Ti-Pure" R-992 | .07 | .004 | <.0004 | .01 | .0025 | .001 |
| "Titanox" RA-NC | .05 | .0025 | <.0004 | .001 | <.001 | .001 |
| "Ti-Pure" R-610 | .02 | .004 | <.0004 | .01 | <.001 | .004 |
| "Titanox" RA-50 | .008 | .0025 | .001 | .004 | .0025 | .001 |

Table 3 (Contd)

| Titanium Dioxide | $\Phi_{rel}$ | Si | Sn | Zn | Zr |
|---|---|---|---|---|---|
| Baker Reagent | 1.0 | .004 | <.001 | <.04 | .004 |
| "Titanox" AMP | .6 | .01 | .001 | <.04 | .004 |
| "Unitane" 0520 | .6 | .04 | <.001 | <.04 | .009 |
| "Titanox" AA | .3 | .01 | <.001 | <.04 | .007 |
| "Ti-Pure" 33 | .2 | .004 | <.001 | <.04 | .004 |
| "Titanox" A-168-LO | .08 | .004 | <.001 | <.04 | .004 |
| "Ti-Pure" R-110 | .1 | .02 | .001 | <.04 | .004 |
| "Ti-Pure" R-100 | .08 | .004 | <.001 | <.04 | .001 |
| "Ti-Pure" R-992 | .07 | 3.0 | .001 | 2.5 | .007 |
| "Titanox" RA-NC | .05 | .01 | <.001 | <.04 | .004 |
| "Ti-Pure" R-610 | .02 | .04 | .001 | 2.5 | .008 |
| "Titanox" RA-50 | .008 | .25 | .001 | .1 | .007 |

(a) Limits of Detection for Metals Not Detected

| Metal | Limit, Wt % | Metal | Limit, Wt % |
|---|---|---|---|
| Ag | .0001 | K | .3 |
| As | .04 | La | .002 |
| B | .001 | Mo | .001 |
| Ba | .0001 | Na | .04 |
| Be | .0004 | P | .2 |
| Bi | .001 | Sb | .01 |
| Cd | .01 | Sr | .0001 |
| Ce | .01 | Th | .001 |
| Co | .004 | U | .04 |
| Ga | .001 | V | .0004 |
| Hg | .04 | W | .04 |
| In | .001 | | |

Effect of Metal Salts on Commercial Pigment Photoactivities.
To determine that the deactivation phenomenon was not restricted
to the reagent anatase, the 12 commercial pigments were coated
with 3 wt % of cobalt as cobalt acetate. A large reduction in
photoactivity was observed for all the pigments evaluated except
the least active pigment, Titanox RA-50 (Table 4). As its activity
was sufficiently low before coating, the change (0.003) is probably
not significant. The deactivation phenomenon is operable on both
anatase and rutile pigment and on both coated and uncoated pigment.

Effect of Metal Concentration on Pigment Photoactivity.
Pigment photoactivity was found to depend on the concentration of
metal coated on the pigment surface (Table 5). Whereas a detectable amount of deactivation of reagent anatase and a moderately
active rutile pigment generally occurred at 0.1 wt % cobalt,
manganese, and cerium levels, large deactivation effects were
observed at 1% metal levels. Increasing the metal concentration
from 1 to 10% generally produced only a small decrease in photoactivity (compared with the change from 0.1 to 1%).

Pigment-Sensitized Dye Photooxidation. Dye degradation
sensitized by titanium dioxide is a recognized phenomenon (16).
As a beginning toward understanding and preventing this phenomenon, the effect of selected amines and phenol stabilizers on the
pigment-sensitized photooxidation of the leuco triphenylmethane
dye I to the blue dye II was determined (Scheme 2).
The hindered amines, tetramethylpiperidin-4-ol and its ester,
as well as the zinc salt of 3,5-di-tert-butylbenzoic acid, effectively inhibited the pigment-sensitized photooxidation of I (Table
6). Commercial ultraviolet stabilizers such as 2,4-di-tert-butyl-
phenyl 3,5-di-tert-butyl-4-hydroxybenzoate, and 2-(3,5-di-tert-
amyl-2-hydroxyphenyl)benzotriazole were ineffective. Similarly,
commercial antioxidants such as octadecyl 3-(3,5-di-tert-butyl-4-
hydroxyphenyl)propionate and 2,6-di-tert-butyl-4-methylphenol
were relatively ineffective. The mechanism of action of the
hindered amines apparently involves radical termination (antioxidant action) (17), but these relatively new compounds are still
the subject of investigation. The failure of the commercial
antioxidants and ultraviolet stabilizers rules out the conventional UV screening and quenching, and primary antioxidant mechanisms
as important factors in inhibiting this pigment-sensitized oxidation process. The high effectiveness of the zinc salt is consistent with the isopropyl alcohol oxidation results, thus showing
that metal salts, including zinc, effectively inhibit oxidations
at the pigment surface.
The effectiveness of zinc is related to its associated anion
(Table 7); the chloride, acetate, or benzoate salts were the
most effective. The maximum effectiveness of the metal would be
expected if the exchange reaction with the pigment surface was

Table 4

Effect of Cobalt on Photoactivity of
Commercial Titanium Dioxide Pigments

| Pigment | $\Phi_{rel}$ | |
|---|---|---|
| | Untreated | Cobalt-Treated[a] |
| Baker Reagent | 1.0 | 0.006 |
| "Titanox" AMP | 0.6 | 0.01 |
| "Unitane" 0520 | 0.6 | 0.01 |
| "Titanox" AA | 0.3 | 0.005 |
| "Ti-Pure" 33 | 0.2 | 0.01 |
| "Titanox" A-168-LO | 0.08 | 0.006 |
| "Ti-Pure" R-110 | 0.1 | 0.03 |
| "Ti-Pure" R-100 | 0.08 | 0.005 |
| "Ti-Pure" R-992 | 0.07 | <0.005 |
| "Titanox" RA-NC | 0.05 | <0.005 |
| "Ti-Pure" R-610 | 0.02 | <0.005 |
| "Titanox" RA-50 | 0.008 | 0.005 |

[a] Surface coated with cobaltous acetate to provide 3 wt % cobalt.

Table 5

Effect of Metal Concentration on Photoactivity of Titanium Dioxide

| Pigment | Metal Conc, Wt % | $\Phi_{rel}$ | | |
|---|---|---|---|---|
| | | $Co^{+2}$ | $Mn^{+2}$ | $Ce^{+3}$ |
| Reagent Anatase | None | 1.0 | 1.0 | 1.0 |
| | 0.1 | 0.5 | 0.3 | 0.5 |
| | 1.0 | 0.06 | 0.01 | 0.05 |
| | 10.0 | 0.02 | <0.005 | 0.01 |
| "Ti-Pure" R-100 | None | 0.08 | 0.08 | 0.08 |
| | 0.1 | 0.1 | 0.003 | 0.02 |
| | 1.0 | 0.02 | <0.005 | 0.01 |
| | 10.0 | 0.01 | <0.005 | 0.005 |

Table 6

Inhibition of $TiO_2$ - Sensitized Photooxidation of a
Leuco Dye by Amines and Phenols

| Additive (1.6 Mol % Based on Reagent Anatase TiO2) | Relative Leuco Dye Oxidation |
|---|---|
| None | 100 |
| 2,2,6,6-Tetramethylpiperidin-4-ol | 2 |
| Di-(2,2,6,6-Tetramethyl-4-piperidinyl) sebacate | 5 |
| 2,4-di-tert-Butylphenyl 3,5-di-tert-butyl-4-hydroxybenzoate | 86 |
| Methyl 3,5-di-tert-butyl-4-hydroxybenzoate | 62 |
| 2-(3,5-di-tert-Amyl-2-hydroxyphenyl)benzotriazole | 62 |
| Octadecyl 3-(3,5-di-tert-butyl-4-hydroxyphenyl)-propionate | 40 |
| 2,6-di-tert-Butyl-4-methylphenol | 30 |
| 3,5-di-tert-Butyl-4-hydroxybenzoic acid | 24 |
| Zinc 3,5-di-tert-Butyl-4-hydroxybenzoate | 1 |

Table 7

Inhibition of $TiO_2$ - Sensitized Photooxidation
of a Leuco Dye by Zinc Salts

| Zinc Salt ($Zn^{+2}/Ti^{+4}$ = 3 Mol %) | Relative Leuco Dye Oxidation[a] |
|---|---|
| None | 100 |
| Zinc formate | 44 |
| Zinc carbonate | 42 |
| Zinc sulfate | 43 |
| Zinc oxalate | 39 |
| Zinc chloride | 8 |
| Zinc acetate | 7 |
| Zinc benzoate | 2 |
| Zinc 3,5-di-tert-butyl-4-hydroxybenzoate | 1 |

[a] Solution contained 0.0025 mmol leuco dye I.

complete (Scheme 3).  Solubility and related factors obviously
influence the rate of this reaction, and no attempts to ensure
complete reaction were made during these experiments.

Effects of Pigments and Stabilizers on Polypropylene Weathering.  Weathering of pigmented and unpigmented polypropylene
formulations was done to confirm the performance of the zinc salts
observed during the isopropyl alcohol and leuco dye oxidation
screening experiments (Table 8).  The weathering lifetime of
polypropylene was increased by a factor of three (from 100 to
300 hr) by incorporating 5 wt % of a moderately active rutile
pigment.  Zinc benzoate, a hindered amine, and a phenolic ultraviolet stabilizer all provide increases in the lifetime of the
pigmented polymer.  Combinations of zinc benzoate or zinc acetate
with the phenolic ultraviolet stabilizer (Items 6 and 7) resulted
in a lifetime of 2500 hr.  An inactive rutile pigment alone was
an effective stabilizer for polypropylene with a hindered phenol
antioxidant (Item 9).  However, incorporating either zinc acetate
alone (Item 11), or zinc acetate or zinc benzoate with a phenolic
ultraviolet stabilizer (Items 12 and 13), provided lifetimes of
3500-4000 hr.  The metal salts obviously function as stabilizers/
deactivators in polymeric media in the same manner as was observed
in the oxidations discussed previously.

A metal probably functions in deactivating titanium dioxide
by participating in the redox cycle at the pigment surface.  The
absorption of a photon results in the generation of a positive
hole in the pigment crystal lattice and electron capture by
adsorbed oxygen to form the superoxide radical anion (Scheme 1).
The chemically bound or adsorbed metal (Me) could transfer an
electron to the pigment to anihilate the positive hole and recap-
ture the electron from superoxide to complete the cycle (Scheme 4).
Whereas this cycle could readily occur with a transition metal
such as manganese, the oxidation of a divalent group IIb metal
such as zinc to a higher oxidation state is highly improbable; the
first, second, and third ionization potentials of zinc are 9.39,
17.89, and 40.0 eV (18).  This redox cycle may be operable with
zinc if the superoxide and zinc are sufficiently close that the
zinc/superoxide complex functions as the electron donor.  However,
an additional explanation based on peroxide decomposition was
sought.

Decomposition of Peroxides by Various Stabilizers.  The
efficiency of tert-butyl hydroperoxide decomposition in tert-butyl
alcohol by various additives was determined (Table 9).  Under
the conditions of these experiments, the phenolic antioxidants
and dilauryl thiodipropionate had little or, often, no effect on
the hydroperoxide decomposition.  The three zinc salts effectively inhibited peroxide decomposition.  This effect might briefly
inhibit the onset of substrate oxidation under weathering-test
conditions, but the peroxide would decompose whenever its concentration reached a sufficient level to permit significant light

Scheme 2

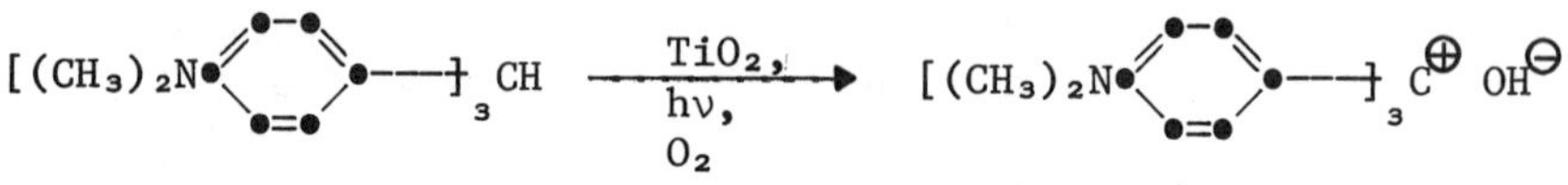

$$[(CH_3)_2N \bullet \bigodot \bullet ]_3 CH \xrightarrow[\substack{h\nu, \\ O_2}]{TiO_2,} [(CH_3)_2N \bullet \bigodot \bullet ]_3 C^{\oplus} OH^{\ominus}$$

|                |                |
| :------------: | :------------: |
| I              | II             |
| Leuco Dye      | Blue Dye       |
| (Colorless)    | (Crystal Violet) |

Scheme 3

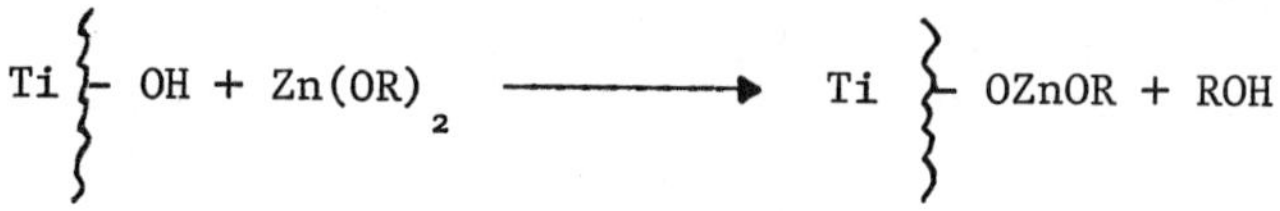

$$Ti \Big\{ OH + Zn(OR)_2 \longrightarrow Ti \Big\} OZnOR + ROH$$

Scheme 4

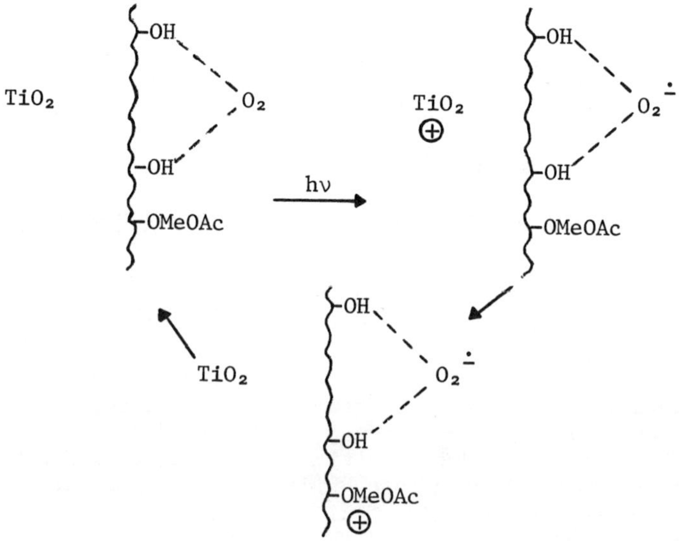

Table 8

Stabilization of Pigmented Polypropylene With Zinc Salts

| Item | Additives,[a] Wt % | Time-to-Embrittlement,[b] Hr |
|------|--------------------|------------------------------|
| 1 | None | 100 |
| 2 | "Ti-Pure" R-100 (5) | 300 |
| 3 | Item 2 + Zinc benzoate (0.5) | 720 |
| 4 | Item 2 + di(2,2,6,6-Tetramethyl-4-piperidinyl) sebacate (0.1) | 2000 |
| 5 | Item 2 + 2,4-di-tert-Butylphenyl 3,5-di-tert-butyl-4-hydroxybenzoate (0.5) | 1400 |
| 6 | Item 5 + Zinc benzoate (0.5) | 2500 |
| 7 | Item 5 + Zinc acetate (0.5) | 2500 |
| 8 | Item 5 + Zinc 3,5-di-tert-butyl-4-hydroxybenzoate (0.5) | 2400 |
| 9 | "Ti-Pure" R-966 (5), Pentaerythrityl tetrakis(3,5-di-tert-butyl-4-hydroxy-phenyl propionate (0.1) | 2000 |
| 10 | Item 9 + Zinc benzoate (0.5) | 2100 |
| 11 | Item 9 + Zinc acetate (0.5) | 3500 |
| 12 | Item 10 + 2,4-di-tert-Butylphenyl 3,5-di-tert-butyl-4-hydroxybenzoate (0.5) | 3800 |
| 13 | Item 11 + 2,4-di-tert-Butylphenyl 3,5-di-tert-butyl-4-hydroxybenzoate (0.5) | 4000 |

[a] To polypropylene melt-processed into 5-mil-thick films.

[b] Samples exposed at 63°C in a "Uvatest" weathering device.

Table 9

Decomposition of t-Butylhydroperoxide in t-Butyl Alcohol

| Additives, mmol | Loss of Hydroperoxide (80°C/24 Hr), % |
|---|---|
| None | 58 |
| Dilauryl 3,3'-thiodipropionate (0.2) | 40 |
| Zinc 3,5-di-tert-butyl-4-hydroxy-benzoate (0.2) | 7 |
| 3,5-di-tert-Butyl-4-hydroxybenzoic acid (0.2) | 31 |
| Zinc benzoate (0.2) | 2 |
| Zinc acetate (0.2) | 2 |
| 3,5-di-tert-Butyl-4-methylphenol (0.2) | 47 |
| Pentaerythrityl tetrakis (3,5-di-tert-butyl-4-hydroxyphenyl)propionate (0.05) | 35 |

Table 10

Decomposition of Hydrogen Peroxide in t-Butyl Alcohol

| Additives, mmol | Loss of Hydrogen Peroxide (80°C/24 Hr), % |
|---|---|
| None | 16 |
| Dilauryl 3,3'-thiodipropionate (0.2) | 56 |
| di(2,2,6,6-Tetramethyl-4-piperidinyl) sebacate (0.1) | 9 |
| Zinc 3,5-di-tert-butyl-4-hydroxybenzoate (0.2) | 85 |
| Zinc benzoate (0.2) | 74 |
| Zinc acetate (0.2) | 70 |
| 3,5-di-tert-Butyl-4-hydroxybenzoic acid (0.2) | 9 |
| 3,5-di-tert-Butyl-4-methylphenol (0.2) | 6 |
| Pentaerythrityl tetrakis(3,5-di-tert-butyl-4-hydroxyphenyl)propionate (0.05) | 7 |

absorption.  Hydrogen peroxide decomposition rates in tert-butyl
alcohol were then determined (Table 10).  The hindered phenols
and hindered amines were inactive as hydrogen peroxide decomposers.  However, dilauryl thiodipropionate and three zinc salts
efficiently decomposed hydrogen peroxide.  Thus, in addition to
hole annihilation, the alternative explanation of hydrogen peroxide
decomposition must also be considered.  Depending on the metal,
nature of the substrate and pigment surface, temperature, and
other rate-influencing factors, one or both inhibition mechanisms may be important.

## Conclusions

Certain metal salts effectively reduce the photoactivity of
titanium dioxide pigments.  Combination of these salts with an
appropriate antioxidant and/or ultraviolet stabilizer provided
highly efficient stabilization of polypropylene.  The deactivation/
stabilization performance of the metal salts is adequately explained on the basis of their decomposition of hydrogen peroxide at the
pigment surface and by annihilation of positive holes in the
pigment crystal lattice.

## References

1.  J. Barksdale, J. L. Turner, and W. W. Plechner, Protective and
    Decorative Coatings, Vol. 2, J. J. Mattiello, editor, Wiley,
    New York, 1942, pp 389-417.
2.  E. Hoffmann and A. Saracz, J. Oil Colour Chem. Assoc., 55,
    1079-1085 (1972).
3.  D. P. Richards and G. W. Bovenizer, J. Paint Technol., 44,
    90-96 (1972).
4.  P. Bentley, J. F. McKellar, and G. O. Phillips, Rev. Prog.
    Color. Relat. Top., 5, 33-48 (1974).
5.  Gether Irick, Jr., J. Appl. Polym. Sci., 16, 2387-2395 (1972).
6.  N. S. Allen, J. F. McKellar, G. O. Phillips, and C. B. Chapman,
    J. Polym. Sci., Polym. Lett., 12, 723-727 (1974).
7.  G. N. Schrauzer and T. D. Guth, J. Am. Chem. Soc., 99, 7189-
    7193 (1977).
8.  R. I. Brickley and V. Venkataraman, Nature, 280, 306-308
    (1979).
9.  B. Kraeutler and A. J. Bard, J. Am. Chem. Soc., 100, 2239-
    2240 (1978).
10. S. P. Pappas and W. Kuhhirt, J. Paint Technol., 47, 42-48
    (1975).
11. C. Naccache, P. Meriaudeau, and M. Che, Trans. Faraday Soc.,
    67, 506-512 (1971).
12. W. A. Weyl and T. Förland, Ind. Eng. Chem., 42, 257-263
    (1950).
13. W. F. Sullivan, Prog. Org. Coat., 1, 157-203 (1972).
14. W. M. Corbett, J.F.L. Roberts, and J. M. Yates, U.S. Patent
    3,547,882 (1970).

15. W. Costain, H. J. Palmer, and T. R. White, U.S. Patent 3,352,821 (1967).
16. C. H. Giles and R. B. McKay, Text. Res. J., _33_, 528–577 (1963).
17. D. W. Grattan, A. H. Reddock, D. J. Carlsson, and D. M. Wiles, J. Polym. Sci., Polym. Lett., _16_, 143–148 (1978).
18. F. A. Cotton and G. Wilkinson, _Advanced Inorganic Chemistry_, 2nd Ed., Interscience, New York, 1966, p 600.

RECEIVED October 20, 1980.

# The Chemical Nature of Chalking in the Presence of Titanium Dioxide Pigments

HANS G. VÖLZ, GUENTHER KAEMPF, HANS GEORG FITZKY, and ALOYS KLAEREN

BAYER AG, D-4150 Krefeld 11 and D-5090 Leverkusen, West Germany

Pigmented paint systems exposed to weathering are liable to undergo an oxidative destruction in the course of time. At an advanced stage of destruction, the pigment particles - originally completely surrounded by binder - are laid bare and can easily be removed from the surface of the paint film. As the effect continues, the pigment particles laid bare in the upper paint layers are progressively washed off by atmospheric precipitation. In the case of white or light pigments one speaks of the phenomenon of "chalking" to express that the pigment particles - like chalk - stick on being touched once they are laid bare by the destruction of the paint matrix. Anatase pigments show a far higher degree of chalking than rutile pigments and are therefore unsuitable for the production of e. g. outdoor paints.

## 1. Chalking process.

**1.1 Chalking.** The destruction of the binder which we call chalking has been known and feared for a very long time. The definition according to ASTM D 659 is as follows:

"Chalking is that phenomenon manifested in paint films by the presence of loose removable powder, evolved from the film itself, at or just beneath the surface. Chalking may be detected by rubbing the film with the fingertip or other means."

Chalking thus involves chemical changes in the binder due to the effect of atmospheric and meteorological influences. Rainfall washes the remaining non-gaseous decomposition products away until the pigment is finally exposed. On being subjected to further weathering, the pigment exposed in the upper layers of paint is finally carried away by the rainfall.

0097-6156/81/0151-0163$05.00/0

Simple as the word "chalking" may sound, the process behind it is extremely complicated. There are in fact two main processes which are responsible for the destruction of the binder as outlined above:
- UV degradation (UVD)
- the photocatalytic oxidation cycle (POC)

UV degradation is a chemical destruction process which leads, under short-wave light, to the direct oxidation of the binder by means of atmospheric oxygen. The reaction takes place mainly via photo-activated states of the binder macromolecules. Excited states of the $O_2$ molecules play a minor role. UVD occurs without the pigment being involved and is therefore of no further interest in this connection.

In the photocatalytic oxidation cycle, on the other hand, the pigment plays a part as a catalyst. The process of photocatalytic oxidation caused by $TiO_2$ pigments is fully understood now (1, 2, 3, 4). It is the subject of the present paper, which comprises a description of this oxidation cycle, the reasons for it as shown by experiments and the practical conclusions.

1.2 Photocatalytic oxidation cycle. The POC is explained below with the aid of Fig. 1 (4).

The starting point is a boundary interface between the $TiO_2$ particles and the binder; on the $TiO_2$ surface, the presence of water has led to the formation of surface hydroxyl groups $[Ti^{4+} \cdot OH^-]$. The first step consists in the absorption of a quantum h$\nu$ of short wavelength and the formation of an exciton (electron/hole pair):

(1)      h$\nu$ $\xrightarrow{[Ti^{4+} \cdot OH^-]}$ e + p

The exciton reacts further instantaneously with a surface hydroxide ion and a $Ti^{4+}$ ion of the lattice, forming a hydroxyl radical and a (formal) $Ti^{3+}$ ion.

(2)      p + $[Ti^{4+} \cdot OH^-] \longrightarrow [Ti^{4+}]$ + $\cdot OH$

(3)      e + $[Ti^{4+}] \longrightarrow [Ti^{3+}]$.

The next step is the addition of an atmospheric oxygen molecule, which takes over the electron from the $Ti^{3+}$ and turns into $O_2^-{}_{(ads)}$:

(4)      $[Ti^{3+}]$ + $O_2 \longrightarrow [Ti^{4+} \cdot O_2^-{}_{(ads)}]$

This is followed by the important reaction in which water is consumed. The water reacts in the form of its dissociation products:

(5)     $O_{2(ads)}^{-}$ + $H^{+}$ ⟶ $HO_{2}^{.}$

(6)     $[Ti^{4+}]$ + $OH^{-}$ ⟶ $[Ti^{4+} .OH^{-}]$.

As a result, we get a perhydroxyl radical. At the same time the $TiO_2$ surface has returned to its initial state and the cycle is completed. The reaction as a whole looks like this:

(7)     $H_2O + O_2 \xrightarrow{\quad TiO_2 , \ hv \quad} {}^{.}OH + HO_{2}^{.}$.

The high reactivity of the radicals ${}^{.}OH$ and $HO_{2}^{.}$ is well known; it brings about the oxidative destruction of the binder, which we call the "photocatalytic oxidation cycle" (POC). It is a cycle in which water and oxygen are constantly being consumed to destroy the binder.

## 2. Experimental arguments.

   **2.1 Influence of water.** The explanation of the POC is the result of studies lasting more than ten years (4). The most important stages and facts are presented below. When we started our work, other authors were only discussing reactions for the POC in which water was not present. (What is meant here is $H_2O$ as a chemical reaction partner, not the "washing-off liquid"). This is surprising, because a number of authors had repeatedly observed that no destruction by POC can take place without water. We therefore began our own studies by devoting particular attention to reactions with water. We carried out comparative measurements of the water vapor and oxygen permeability of binder films. The result we obtained was that the permeability of water through pigmented binders under practical conditions is on average $10^3$ greater (calculated on the number of mols) than that of $O_2$. (Tab. I) This showed good agreement with data given in literature on unpigmented binders and polymers.
   In a special weathering device we constructed ourselves (4), we later carried out chalking tests in which great care was taken to exclude water while allowing air to enter unhindered. This and the

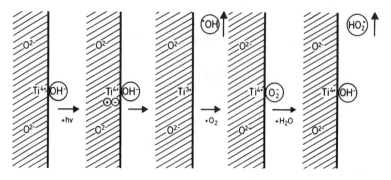

Organic Coatings and Plastics Chemistry

*Figure 1.   Schematic of the photocatalytic oxidation cycle (POC) (12)*

**Table I.   Permeation of $H_2O$ and $O_2$ Through Paint Films
Under Atmospheric Conditions (2)**

|  | Molar permeation coefficient of $O_2$ | Molar permeation coefficient of $H_2O$ | Ratio of molar coefficients of $O_2$ and $H_2O$ |
|---|---|---|---|
| Desmodur/Desmophen paint, unpigmented | $3{,}1 \cdot 10^{-11}$ | $2{,}0 \cdot 10^{-7}$ | $6{,}0 \cdot 10^{3}$ |
| Desmodur/Desmophen paint, 8,5% p.v.c., rutile untreated | $3{,}6 \cdot 10^{-11}$ | $1{,}0 \cdot 10^{-7}$ | $3{,}0 \cdot 10^{3}$ |
| Nitrosynthetic lacquer, unpigmented | $8{,}0 \cdot 10^{-11}$ | $4{,}5 \cdot 10^{-8}$ | $0{,}6 \cdot 10^{3}$ |
| Nitrosynthetic lacquer, 8,5% p.v.c., rutile untreated | $8{,}0 \cdot 10^{-11}$ | $4{,}3 \cdot 10^{-8}$ | $0{,}5 \cdot 10^{3}$ |
| Alkyd enamel, unpigmented | $9{,}3 \cdot 10^{-11}$ | $4{,}5 \cdot 10^{-8}$ | $0{,}5 \cdot 10^{3}$ |
| Alkyd enamel, 8,5% p.v.c., rutile untreated | $10{,}0 \cdot 10^{-11}$ | $4{,}0 \cdot 10^{-8}$ | $0{,}4 \cdot 10^{3}$ |

Farbe & Lack

subsequent tests were carried out with air-drying
binders based on alkyd resins and also with plastics
(5). In order to prevent UVD as far as possible,
suitable cut-off filters were used. To record the
level of destruction, the weight loss of the
specimens was measured in relation to the weathering
time (gravimetric test). A series of paint specimens
with graduated pigment volume concentrations (p.v.c.)
was placed in the weathering device. The result was
that, despite the presence of atmospheric oxygen,
the POC actually comes to a complete standstill when
$H_2O$ is absent (Fig. 2). It is the reactions (5) and
(6) which cannot take place - the cycle is interrupted after equation (4).

As has been shown, the POC starts from surface
hydroxyl groups which are constantly being newly
formed in the cycle by $H_2O$. The existence of these
hydroxyl groups was insured when we began our
investigations; in Germany, it had become known
particularly as a result of the studies carried out
by BOEHM at al. (6), which were based on the research
results of other authors.
    2.2 Influence of atmospheric oxygen. At the
time we started our studies, none of the other
authors had satisfactorily dealt with the question
of whether oxygen as well as water takes part in
the POC.

Our own tests on this question were carried out
in the previously mentioned special weathering device
(4). This device offers the possibility of producing
weathering conditions with an exactly pre-determined
gas phase. Paint specimens with varying p.v.c. of
$TiO_2$ pigment (anatase and rutile) were weathered
with irrigation, but making sure to exclude any
oxygen. Here, too, the POC came to a complete standstill (Fig. 2).

Weathering under these conditions led to a
marked graying of the specimens, which declined
slowly when they were placed in air again. If the
grayed samples were put into hot water, the graying
receded within a few seconds (Fig. 3). The explanation is that the POC is stopped before the reaction
(4); the gray shade is the visible sign of the
presence of $Ti^{3+}$. Only a massive supply of $H_2O$ will
ensure that the reactions of equations (4) and (5)
continue unhindered.

These tests were also repeated with the
simultaneous exclusion of $H_2O$ and $O_2$. As expected,
there is no photocatalytic oxidation and the
specimens do not turn gray. We naturally also carried

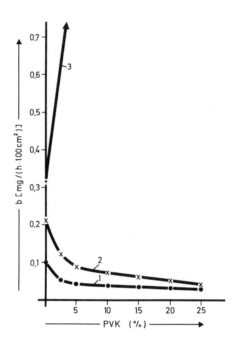

Farbe & Lack

Figure 2. Rate of degradation = f(PVC); anatase/alkyd resin: (1) weathering under exclusion of $H_2O$; (2) weathering under exclusion of $O_2$; (3) weathering in the presence of $H_2O$ and $O_2$ (4)

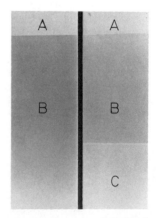

Figure 3. Regeneration of the graying by water: (A) array covered against irradiation; (B) irradiated array; (C) regenerated array by dipping in water (4)

Farbe & Lack

out tests in the presence of $H_2O$ and $O_2$ parallel to
all of the series and always observed destruction
through the POC (Fig. 2). These tests are an impressive demonstration of the interplay of $H_2O$ and $O_2$ in
the photocatalytic oxidation cycle.

2.3 Dependence on the wavelength of the incident
light. Knowing the threshold wavelength $\lambda_g$, above
which no further POC destruction takes place, is of
interest for several reasons. On the one hand, it
was expected that the value of $\lambda_g$ is related to the
energetic data of the rutile and anatase lattice.
Secondly, it is possible on knowing $\lambda_g$ to separate
the part of the UVD in the chalking being undesirable
for testing the photochemical pigment activity from
the POC part. We made use in our tests of UV cut-off
filters, which cut off the irradiation spectrum
below $\lambda = 300, 335, 375, 395, 435$ nm (7). The most
important result we obtained was that the highest
relative destruction rate is obtained on coatings
pigmented with $TiO_2$ anatase using the 375 nm cut-off
filter (related in each case to the UVD of the unpigmented film). Shifting the absorption edge in the
direction of higher wavelengths, the influence of
the pigment declines in leaps and bounds and the
dependence on the PVC disappears (Fig. 4). This
proves that $\lambda_g$ must be between 375 and 395 nm for
anatase. Below 375 nm, the UVD becomes increasingly
responsible. This finding harmonizes very well with
the concept that the primary step in the POC is the
formation of an exciton. $TiO_2$ is an extrinsic semiconductor of the n-type and the distance between the
valence and the conduction band is:
- for rutile, 3.05 eV, i.e. optical absorption edge
  415 nm
- for anatase, 3.29 eV, " "     "          "        "
  385 nm.

With the absorption of a quantum with an energy
of more than 3.05 eV resp. 3.29 eV, an electron is
lifted out of the valence band and into the conduction band, thereby forming an exciton (Fig. 5). This
interpretation is also supported by the molecular
orbital theory and the crystal field theory regarding the bonding conditions in the $TiO_2$ lattice.

2.4 ESR determination of radical species.

2.4.1 OH and $HO_2$ radicals. Radical states are
generally connected with the presence of unpaired
electrons, which makes the ESR method suitable for
determining them. ESR determination of the OH radical
in the POC was published by us in 1970 (1). With the
aid of test spectra, two lines $g_1 = 2.0118$ and

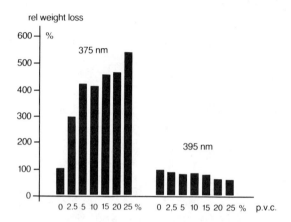

N.V.V.T.

*Figure 4.   Total weight loss dependence of anatase pigmented paint films on wave-*
*length and on p.v.c. (7)*

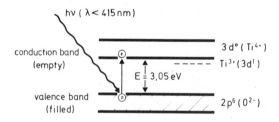

Farbe & Lack

*Figure 5.   Schematic of energy band model of the rutile form of TiO$_2$ (4)*

$g_2 = 2.0134$ were observed between $-60\ ^{\circ}C$ and $-175\ ^{\circ}C$. According to the state of the knowledge at that time, $g_1$ was attributed to the free radical and $g_2$ to a complex compound. Nowadays, the view is that both lines belong to complex compounds from $Ti^{4+}$ $^{\cdot}OH$, $HO_2^{\cdot}$ and $H_2O_2$. A decision as to which line should be attributed to the $^{\cdot}OH$ or the $HO_2^{\cdot}$ complex is so difficult because both species are in correlation to each other according to

(8) $\qquad HO_2^{\cdot} + H_2O \xrightleftharpoons{\hspace{1cm}} {}^{\cdot}OH + H_2O_2$

Many authors have carried out a great deal of work on proving the existence of $H_2O_2$. For the first time PAPPAS and FISCHER suceeded in determining it with the aid of an extremely sensitive analysis method with peroxidase as used in biochemistry (8). The concentrations determined in the POC are extremely low and are a "by-product" of the POC, but are of no further importance within the POC itself. If $H_2O_2$ were the oxidizing agent in the POC, then photoinactive substances like ZnO or CdS ought to show a very high destruction by POC since in aqueous systems they provide $H_2O_2$ concentrations which are several orders of magnitude higher.

It was found time and again in these studies that anatase, rutile and treated rutile pigments differed in the quantity of radicals they produced (Fig. 6-8). These differences correspond with the familiar stages in the photochemical activity of these pigments.

2.4.2 $Ti^{3+}$ centers. It was also possible to demonstrate the formation of paramagnetic $Ti^{3+}$ centers during the weathering of $TiO_2$ with the aid of our ESR measurements at $1.3\ ^{\circ}K$. Here, too, anatase and rutile showed quantitative differences in conformity with the differences in the photoactivity of both modifications (1).

2.4.3 Adsorbed $O_2^-$. Several authors have observed that $O_{2(ads)}^-$ is formed on the $TiO_2$ surface under chalking conditions. We were able to prove this in our ESR measurements (4). At $g = 2.009$, two ESR lines indicate the transfer of the electron to the added $O_2$ molecule.

2.5 Simulation of destruction by POC. The description of the POC given here leads us to the idea of "simulating" the cycle, i.e. specifically producing OH or $HO_2$ radicals and allowing them to react with the paint film. To produce the radicals, we made use of the decomposition of water molecules

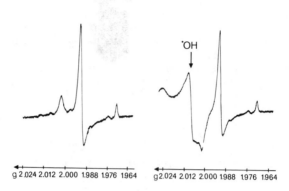

*Figure 6. ESR measurements on anatase/water: (left) not exposed to light; (right)
exposed to light*

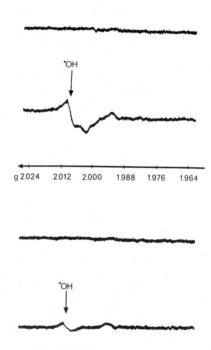

*Figure 7. ESR measurements on un-
treated rutile TiO₂ in water: (top) not
exposed to light; (bottom) exposed to light*

*Figure 8. ESR measurements on treated
rutile/water: (top) not exposed to light;
(bottom) exposed to light*

by high-frequency discharge (Fig. 9) within a
specially developed apparatus. The tests were carried
out with paint specimens pigmented with TiO$_2$ (anatase
and rutile), and parallel to this, the same specimens
were subjected to normal weathering in a conventional
device. When distinct signs of chalking appeared,
tests were carried out on both series of specimens
using the KEMPF test. We found that the replicas of
the specimens from the high-frequency apparatus did
not differ from those from the weathering device (1).
These results are in an impressive agreement with
the fact that hydroxyl and perhydroxyl radicals are
responsible for the phenomena which we observed in
the POC and which contribute to chalking.

### 3. Conclusions, practical consequences.

#### 3.1 Phenomenological consequences.
##### 3.1.1 Photoactivity of TiO$_2$ modifications.
We
should be able to expect of a reaction system drawn
up by experiment that it reflects or explains the
observed phenomena correctly. In part 2, a few such
facts were dealt with in the light of the POC
reaction system. Here are a few more:
- the differing photoactivity of the TiO$_2$ modifications
- the "internal destruction" through the POC
- the protective effect of the TiO2 pigments against
UV degradation.
   Of the modifications anatase and rutile, the
anatase ought to have the highest photostability if
we look only at the position of the absorption edge.
But since the energy content of the exciton is higher
with the anatase than it is with the rutile, the
probability of reaction with the Ti$^{4+}$ ion is greater
despite the smaller number of excitons. The fact that
rutile is more stable is also explained by the more
rigid bond of its surface hydroxyl groups. BOEHM (9)
has shown this by considering the bond strength of the
OH$^-$ on TiO$_2$ fracture surfaces. Accordingly, although
the rutile more readily forms excitons, the
positive hole reacts slower with the OH$^-$.
##### 3.1.2 Internal destruction and volume shrinkage.
Since destruction by the POC takes place at the
interface between TiO$_2$ and binder, therefore binder
is not only decomposed at the surface of the paint
film. There is also a certain amount of destruction
inside and, with sufficient flexibility of the binder
and sufficient adhesion between the pigment and the
binder, this must lead to a shrinkage in volume.

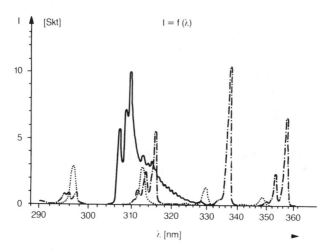

Farbe & Lack

*Figure 9.* *UV emission spectra of various molecules in the high-frequency discharge (2).* $I(\lambda)$ *= intensity, measured on radiation receiver,* $\lambda$ *= wavelength.* (———) OH(H₂O); ( · · · ) O₂; (— · — · ) N₂, *tenfold reduced*

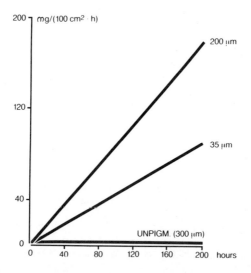

N.V.V.T.

*Figure 10.* *Dependence of degradation of pigmented PE sheets (anatase, 0.2% p.v.c.) of different thickness on weathering time, by gravimetric measurement (7)*

We were able to prove this by several methods
independent of one another:
- Destruction by photocatalytic oxidation in the case
of varying thicknesses of the paint film.
Only when a certain film thickness is reached, do
we find a constant value for the weight loss. If
the film thickness is too low, the radiation is
not used entirely for the POC, some of it passes
unused through the paint film (Fig. 10). If the
destruction would only take place on the surface
of the film, then this effect would not occur.
- Difference between the measured decrease in film
thickness and that calculated from the weight loss.
Since with internal destruction it is only "light"
binder matrix which disappears and not "heavy"
pigment, there must be a difference between the
two measurements. This is also actually found
(Fig. 11).
- Increase in the p.v.c. If there is a shrinkage in
the volume, then this should become evident by an
increase of the p.v.c. Determinations have shown
that the p.v.c. increases by several % (Fig. 12).
    3.1.3 Protective properties of rutile pigments.
A further consequence of our photocatalytic oxidation
cycle is the protective function which rutile pigments exert against UV degradation. This effect can
be shown by several methods:
- Morphological investigations. With the aid of
scanning electron micrographs, it is possible to
demonstrate this effect visibly (10). The test
results can be roughly outlined as follows: with
anatase pigments, destruction by photocatalytic
oxidation takes place more rapidly than by UV
degradation and therefore governs the overall
picture of chalking (Fig. 13). After weathering,
the anatase particles remain on the paint surface
in what might be described as "pits" (Fig. 14).
In the case of treated or coated rutile pigments,
on the other hand, destruction through the POC is
slower than by UV degradation, which means that UV
degradation plays the major role (Fig. 13). Rutile
pigment particles therefore stand on "pedestals",
which have remained behind because the pigment
particles cast a shadow and prevented the UV
radiation from reaching the paint at these regions
(Fig. 15).
- Splitting up the total amount of degraded paint
material into a percentage of pigment and a percentage of binder. Gravimetric analysis of the
destruction process enables the mass loss to be

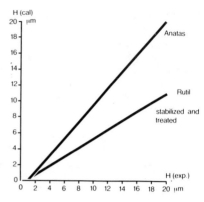

*Figure 11. Comparison between film thickness values H, obtained by experiment and by gravimetric measurement (7)*

**N.V.V.T.**

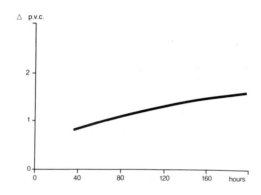

**N.V.V.T.**

*Figure 12. Dependence of increase in p.v.c. of PE sheets on weathering time ( anatase, 15% p.v.c.; 240 μm thickness (7)*

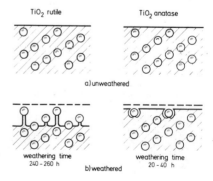

*Figure 13. Schematic of the degradation processes during the weathering of binders pigmented with TiO₂ (2)*

Farbe & Lack

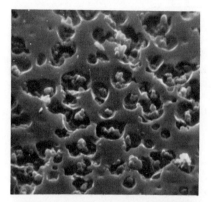

Farbe & Lack

*Figure 14. TiO$_2$ anatase pigmented coating, chalk rating 4, general view (2)*

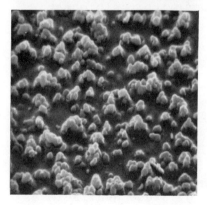

Farbe & Lack

*Figure 15. TiO$_2$ rutile pigmented coating, chalk rating 4, general view (2)*

divided into pigment and binder parts (7). When we look at the proportion of binder that has become destroyed, we find that, with increasing p.v.c.,
- in the case of anatase it also increases (Fig. 16),
- in the case of untreated rutile it remains about the same (Fig. 17),
- in the case of treated rutile it decreases (Fig.18).

The latter observation is proof of the protective effect of stabilized rutile pigments.

### 3.2 Means of reducing chalking.

### 3.2.1 Choosing the most stable $TiO_2$ modification.

It has been shown that chalking is the result of two main processes: UV degradation and the photocatalytic oxidation cycle. What can be done to stabilize $TiO_2$ pigments in such a way that they cause as little chalking as possible? A few possibilities have already been mentioned:
- protection from UV radiation,
- exclusion of water,
- exclusion of oxygen.

It is obvious that these possibilities are unrealistic in practice as a means of preventing chalking. Outdoor coatings can neither be protected from radiation, nor from water, nor from oxygen. There are, however, other possibilities, such as selecting the $TiO_2$ modifications with the highest photostability. This has   already been dealt with. Rutile pigments have a higher stability than anatase pigments, but it can be improved even further by incorporating foreign ions and by suitable after-treatment:

### 3.2.2 Reducing the number of surface Ti-OH groups.

Improved stability is possible by incorporating foreign ions instead of $Ti^{4+}$ into layers near to the surface of the pigment particle. Pigment producers have been doing this for many years. Although the incorporation of $Zn^{2+}$ or $Al^{3+}$ ions into the $TiO_2$ also leads to the formation of surface hydroxyl ions, they are unable to form OH radicals, since there is no suitable energy stage available as with $Ti^{4+}/Ti^{3+}$.

### 3.2.3 Destroying  generated OH and $HO_2$ radicals.

The treated pigment particle is surrounded by a "catalytically active wall" in the form of substances with a very high surface area. The generated OH radicals then react to a large extent according to

$$(9) \qquad 2 \; ^{\cdot}OH \longrightarrow H_2O + \frac{1}{2} O_2$$

and destroy each other by recombination. Use has also

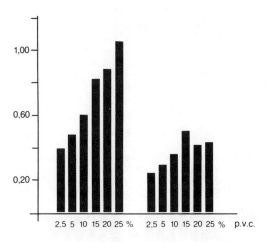

N.V.V.T

*Figure 16. Dependence of total degradation and degradation of binder on p.v.c.; sample: alkyd resin pigmented with TiO$_2$ anatase ((weight loss rate: mg/(100 cm$^2$ · h)) (7)*

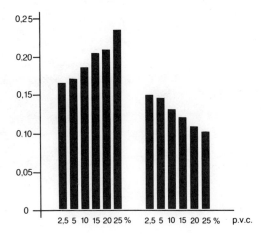

N.V.V.T

*Figure 17. Dependence of total degradation and degradation of binder on p.v.c.; sample: alkyd resin pigmented with TiO$_2$ rutile (untreated) (weight loss rate: mg/ (100 cm$^2$ · h)) (7)*

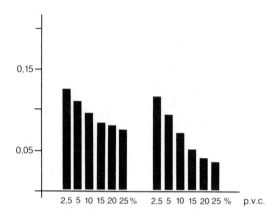

N.V.V.T

*Figure 18.    Dependence of total degradation and degradation of binder on p.v.c.
sample: alkyd resin, TiO₂ rutile (coated) (weight loss rate: mg/(100 cm² · h)) (7)*

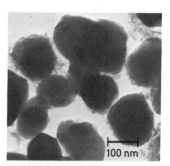

*Figure 19.    TiO₂ rutile pigment treated
with Al and Si aquates (11)*

Farbe & Lack

been made of this mechanism for a long time: The highly stabilized $TiO_2$ rutile pigments are "after-treated" or "coated", usually with one or several layers of Si and Al oxide aqueous compounds. The fact that these compounds have a structure with high surface area (Fig. 19) has already been shown (11). It ensures a fast recombination of the majority of the radicals.

## 4. Conclusions.

The chalking of paints is brought about by two elementary processes: UV-degradation (UVD) and the photocatalytic oxidation cycle (POC). Only with the POC do pigments act as a catalyst. For $TiO_2$ pigments in binders, the photocatalytic oxidation cycle is clear. One water molecule and one oxygen molecule are converted into one hydroxyl and one perhydroxyl radical. The radicals are the cause of binder destruction. Some of the experimental principles for the chemical nature of the POC are dealt with here: the part played by the water, the part played by the oxygen, the dependence on wavelength, ESR determination of the radical species and the simulation of degradation due to the POC. The practical consequences from the reaction system of the POC provide explanations of the differing photoactivity of the $TiO_2$ modifications, an interpretation of the internal destruction combined with volume shrinkage of the paint film and explanations of the protection afforded by rutile pigments against UV-degradation. Measures for reducing chalking follow from the reaction system: the choice of the most stable $TiO_2$ modification (rutile), the reduction of surface Ti-OH through the incorporation of foreign ions into the $TiO_2$ lattice and the destruction of formed OH radicals by suitable aftertreatment of the $TiO_2$ pigment.

Abstract. The photocatalytic oxidation cycle (POC) is that process in chalking in which the pigment participates. This paper has shown the chemical reaction scheme for the course of this process as well as the experimental results confirming this scheme. The practical consequences of the reaction course are to be considered under two different aspects: on the one hand interpretation of the phenomenological observations (like differing photoactivities of the $TiO_2$ modifications, "internal" destruction on chalking, protective function of the rutile pigment against UV degradation), on the other hand the measures that can be taken to reduce the photoactivity of the $TiO_2$ pigments (selection of the modification of the highest stability, incorporation of foreign ions in the $TiO_2$ crystal lattice, surrounding of pigment particles by aftertreatment).

Acknowledgment. We would like to thank Prof. Dr. H.P. BOEHM, Munich University, for contributing many fruitful discussions and interesting suggestions.

Literature Cited

1. Völz, H.G., Kämpf, G., Fitzky, H.G., "X. FATIPEC-Congress, Congressbook", Vlg. Chemie, Weinheim, 1970, p. 107
2. Völz, H.G., Kämpf, G., Fitzky, H.G., farbe + lack, 1972, 78, p. 1037
3. Völz, H.G., Kämpf, G., Fitzky, H.G., Progr. Org. Coat., 1973, 1, p.1
4. Völz, H.G., Kämpf, G., Klaeren, A., farbe + lack, 1976, 82, p. 805
5. Kämpf, G., J. Coat. Technol., 1979, 51, p. 51
6. Boehm, H.P., Herrmann, M., Kaluza, U., Z. Anorg. Allgem. Chemie, 1970, 372, p. 308
7. Völz, H.G., Kämpf, G., Klaeren, A., "XV. FATIPEC-Congress, Congressbook", N.V.V.T., Amsterdam, 1980, p. III-41
8. Pappas, S.P., Fischer, R.N., J. Paint Technol., 1974, 46, p. 65
9. Boehm, H.P., Z. Anorg. Allgem. Chemie, 1967, 352, p. 156
10. Kämpf, G., Papenroth, W., Holm, R., J. Paint Technol., 1974, 46, p. 56
11. Kämpf, G., Völz, H.G., farbe + lack, 1968, 74, p. 37

RECEIVED October 20, 1980.

# Photochemical Studies of Methacrylate Coatings for the Conservation of Museum Objects

ROBERT L. FELLER, MARY CURRAN, and CATHERINE BAILIE

Center on the Materials of the Artist and Conservator, Mellon Institute, Carnegie–Mellon University, Pittsburgh, PA 15213

Methyl and ethyl methacrylate polymers, although extensively used in industry, do not possess the solubility characteristics (low polarity) that would make them appropriate for use over traditional oil paintings and other organic-based museum objects that might be sensitive to polar solvents such as alcohols, ketones and esters. Poly(n-butyl methacrylate), offered as an artists' varnish in the late 1930's, did not become widely accepted in the war-disrupted decade that followed. Accordingly, early in 1951, our laboratory began a detailed study of the higher alkyl methacrylate polymers for potential use as picture varnishes (1).

To be suitable for use in museum preservation and restoration practice, picture varnishes must have a number of defined characteristics: (a) they must not interact with the object in a detrimental way - they must not, for example, soften or dissolve an old painting; (b) they must be easily applied by brushing or spraying; (c) they must have appropriate flexibility and hardness; (d) they must undergo little change in appearance over long periods of time; and (e) they must remain soluble for many generations in solvents of reasonably low polarity so that they can be removed at a later date with little risk to the substrate. We began our search for thermoplastic materials that would fulfill these criteria by investigating the variation in the properties of methacrylate polymers as the size of the alkyl group of the alcohol radical was increased from methyl through hexyl (2,3).

During our initial accelerated-aging tests in a carbon-arc Fade-ometer®, we found that the higher alkyl methacrylates exhibited a tendency to crosslink. Without discoloring - without changing in appearance - a picture varnish made with these initially linear polymers could become insoluble and unswellable in toluene and acetone; potentially it would be impossible to remove such a coating in the future without considerable risk to the painting. This was a serious matter that warranted thorough investigation.

The purpose of the studies that will be reported here was to determine the influence of (a) temperature, (b) wavelength of radiation, (c) intensity of radiation, and (d) structure of the alkyl group upon the tendency of a number of the polymers to crosslink, as evidenced by the rate of formation of insoluble matter.

## Preparation of Polymers

Both commercial and laboratory-synthesized polymers were used. Those made in the laboratory were generally prepared by solution polymerization, refluxing commercially available monomers in toluene using benzoyl peroxide as the catalyst. Other preparations were made in which azo-bis-isobutyronitrile (AIBN) was used as initiator, ethanol was employed as the refluxing medium, and monomers were especially synthesized in the laboratory. These variations in preparative procedure did not significantly affect the ranking of the polymers with respect to their tendency to crosslink, as reported in Table I.

Films of the polymers were generally cast with a drawdown bar from 20% solutions in reagent-grade toluene. All the data reported in the accompanying tables and graphs are based on films thus coated on aluminum foil. The use of the foil speeded up the formation of insoluble matter 1.9 times over that of coatings on glass and also permitted convenient measurements of changes in weight upon exposure and extraction (4). Ordinary 0.001"-thick household aluminum foil was rubbed flat on the surface of glass plates, using a few drops of acetone to aid in the adherence of the foil to the glass and to wet the cotton swabs used to clean the foils and rub them flat. After standing for 24 hours, the cast films were baked 48 hours at 70°C. The retained solvent after baking was usually no more than a few tenths of a percent by weight of resin. The small amount of toluene that remained was considered insufficient to alter the conclusions. Tests have shown, however, that the retention of a chemically active solvent such as turpentine in the film can shorten the induction time considerably (5).

Major differences in molecular weight can be expected to influence the radiation dose necessary to give rise to insoluble matter; because of this, intrinsic viscosity values are given in Table I to indicate that the laboratory-prepared polymers were generally similar in this respect.

## Crosslinking

Influence of Wavelength of Radiation. Our initial indications of crosslinking were observed during exposures in an Atlas Electric Devices Co. carbon-arc Fade-ometer® equipped with a Corex D filter. Later, the same behavior was found when exposures were made in an Atlas 600WRC xenon-arc Fade-ometer® having Pyrex-glass filters.

When crosslinking was first encountered, a number of colleagues advised us that the phenomenon was not likely to be of importance in a museum; there, the natural or fluorescent illumination would consist only of those wavelengths that would pass through ordinary glass. We soon found, however, that, although

Table I: Tendency of Methacrylate Polymers to Become Insoluble Samples on Aluminum Foil, Single-Carbon-Arc Fade-ometer® (Corex D Filter), 62°C

| Alkyl Radical, R, in Poly (R-Methacrylate) | Reactive Hydrogen on Alkyl Carbon Atom No. | Intrinsic Viscosity in MEK, 23°C | Hours to Attain 50% Insolubility |
|---|---|---|---|
| 3-methylbutyl | 3t (a) | .27 | 4 |
| 1:3 p-methylcyclohexyl/isoamyl* | 3,4t | .22 | 35 |
| 2-methylbutyl | 2t | .27 | 49 |
| n-butyl | 3,2s | .25 | 91 |
| 6:4 neopentyl/n-amyl | 4,3,2s | .41 | 110 |
| isobutyl (Elvacite® 2045) | 2t | .59 | 126 |
| n-propyl | 2s | .21 | 660 |
| ethyl (Elvacite® 2042) | 1s | 1.6 | >> 1800 (b) |
| ~7:3 ethylmethacrylate/methyl-acrylate (Acryloid® B-72) | 1s | .20 | >> 2023 (b) |

(a) 3t = tertiary hydrogen on third carbon from alcoholic oxygen; s = secondary hydrogen.

(b) no more than 5% insoluble matter detected.

* Polymer 59 in Figure 3.

"Elvacite" is a registered trademark of the duPont Company, "Acryloid", a registered trademark of the Rohm and Haas Company and "Fade-ometer", a registered trademark of the Atlas Electric Devices Company.

the reaction would be very slow under museum conditions, it was indeed activated by the visible as well as by the near-ultraviolet radiation. For example, when an ultraviolet filter was introduced that removed most of the radiation below 400 nm, crosslinking still took place, but the rate of formation of insoluble material in the carbon-arc Fade-ometer® was reduced to about one-half. The rate of crosslinking was eventually shown to vary almost logarithmically with the shortest wavelength in the illumination, as illustrated in Figure 1.(6)   It is interesting to note in Figure 1 that a number of photochemical degradations follow much the same relationship: loss of weight from alkyd paint films(7), degradation of low-grade paper(8), evolution of hydrogen from rubber(9) and development of carbonyl groups in poly(vinyl chloride) (10). The reason for this particular sensitivity to wavelength is perhaps the absorption of radiation by functional groups present at very low concentrations.

Later investigations have fully confirmed the conclusion that the crosslinking of the higher alkyl methacrylate polymers will take place on a gallery wall. We have been able to demonstrate that an induction time of about 11 years occurs before insoluble matter begins to form in commercial normal and isobutyl polymers on a well-illuminated museum wall (1,11). Protection against such loss of solubility is one of a number of reasons for recommending the use of ultraviolet filters over windows and over fluorescent-lamp sources in museums (12).

The relationship illustrated in Figure 1 should be extended below 313 nm only with caution.   The shorter wavelengths may induce photolytic decomposition, resulting in much more chain scission. Thus, when samples of poly(n-butyl methacrylate) were exposed to a high-pressure mercury lamp (emitting intense radiation at 254 nm), we found that, although gel formation took place in the beginning, considerable chain breaking and volatilization eventually occurred (13,14). Under this lamp, the films developed bubbles and blisters at an early stage, perhaps owing to the formation of monomer by unzipping reactions.   We have warned colleagues not to attempt to use lamps that emit 254 nm radiation in "accelerated-aging" tests of museum materials, because the photoactivated mechanism of deterioration may be distinctly different from that caused by the near ultraviolet and visible radiation.

For extensive studies of the effects of 254 nm radiation on acrylates and methacrylates, the reader is referred to the work of Morimoto and Suzuki(14) and of Grassie (15,16). In 1964, Oster reported on the crosslinking of these and other polymers by the near ultraviolet, sensitized by the presence of 2-methylanthraquinone (17).

Influence of Alkyl Group.  Colleagues and manufacturers further advised us in 1952 that acrylics would not undergo crosslinking because the polymers do not obviously absorb in the near ultraviolet.  (However, the field now realizes that trace components at very low levels may absorb radiation in this range.)  Our advisors

may be forgiven, for their greatest familiarity at the time was
with polymers of methyl and ethyl methacrylate.  As the data in Table I show, we found the tendency-to-crosslink to reside primarily
in the butyl and amyl esters that contain tertiary hydrogens, and
secondarily, in the n-amyl, n-butyl and n-propyl esters which have
secondary hydrogens removed from the backbone of the main chain by
the carboxylic carbon and oxygen and by two alkyl carbons (1,18).
The normal alkyl polymers that we prepared may have contained
traces of an isopropyl, isobutyl or isoamyl impurity; nonetheless,
the presence of tertiary hydrogens is not necessary to account for
the tendency to crosslink.  A six-membered-ring intermediate structure, such as proposed by Grassie(15) and others, may accelerate
the loss of hydrogen atoms from the alkyl groups.  If so, then the
normal butyl, amyl and higher esters should tend to crosslink more
readily than polymers made with the methyl and ethyl esters.  Thus,
Barton(19), who induced crosslinking in n-butyl and n-nonyl methacrylate polymers through the addition of dicumyl peroxide, clearly
demonstrated that the nonyl ester had a higher crosslinking efficiency than the polymer with the shorter side chain.

If the loss of solubility of these initially linear polymers
takes place through a free-radical-chain process in which the crosslinking reaction represents the termination step, one may be hesi-
tant at first to explain the fact that a free radical, generated at
one point on a relatively sluggish polymer chain, can find a radical on a neighboring chain with which to terminate.  However, per-
haps reactions of the type $R\cdot + R'H \longrightarrow RH + R'\cdot$ or $ROO\cdot + R'H \longrightarrow$
$ROOH + R'\cdot$ can take place rapidly between pendant alkyl groups a-
long the chain until a radical is encountered on a neighboring
chain; this would explain in part the enhancement of crosslinking
caused by the tertiary hydrogens being located increasingly further
from the main chain backbone (Table I; Figure 2).  Grassie, Semenov
(20), Chien(21), and others, have proposed such intramolecular reactions by pendant groups along the main chain.

Influence of Intensity of Illumination.  There is a strong
tendency for those who employ intense sources of illumination in
accelerated photochemical aging tests to make quick calculations on
the basis of the reciprocity principle - that is, to assume that,
as the intensity is lowered, the time for equivalent damage will be
correspondingly lengthened.  For a number of reasons, the reciprocity principle need not hold true, and it is advisable to check
this point.  We did so by exposing a readily crosslinkable polymer
(duPont's Elvacite® 2046, a copolymer of normal and isobutyl methacrylate) under a series of intensity-reducing screens.  The experiments were carried out in the xenon-arc Fade-ometer® with Pyrex-
glass filters, in which the circulating air was maintained at 32°C
and about 25% relative humidity.  Aluminum wire screens were employed to decrease the intensity without seriously altering the
spectral distribution of the illumination.  The results, given in
Table II, indicate that the reciprocity law held true over a 30-
fold range of intensity.

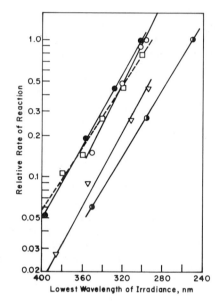

*Figure 1.   Data of various investigators on the influence of the lowest wavelength of irradiance on the rate of a variety of chemical reactions ((●) Feller (6); (◐) Miller (7); (□) Harrison (8); (○) Bateman (9); (▽) Martin/Tilley (10))*

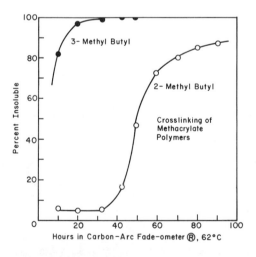

*Figure 2.   Comparison of the rate of formation of insoluble matter in polymers based on 3-methyl- and 2-methyl-butyl methacrylate*

Table II: Check on Reciprocity Law: Induction Time for Cross-linking of Elvacite® 2046 Under Various Intensities of Illumination in a Xenon-Arc Fade-ometer® With Pyrex-Glass Filter

| Relative Intensity, I | Induction Time, $T_i$ (Hours) | $IT_i$ |
|---|---|---|
| 1.00 | 15 | 15.0 |
| .71 | 22 | 15.6 |
| .49 | 29 | 14.2 |
| .35 | 40 | 14.0 |
| .135 | 79 | 10.7 |
| .033 | 530 | 17.5 |

$$14.5 \pm 2.2$$

Further experiments were conducted under the relatively mild conditions of exposure to a bank of "daylight" fluorescent lamps in a room maintained at 70°F (21.1°C) and 50% relative humidity. Under these conditions, the intensity of the near-ultraviolet radiation amounted to 2.5% of the energy in the visible range and the temperature of the samples reached no more than 26°C. Such experiments clearly demonstrated that commercial normal and isobutyl methacrylate polymers, as well as those that are prepared in the laboratory by solution polymerization, develop 50 to 80% insoluble matter after about 9 million footcandle hours of exposure to "daylight" fluorescent lamps. By extrapolation, one may estimate that these coatings would develop the same degree of insolubility in about 50 years of exposure in a daylighted museum under an average level of illumination of about 50 footcandles (6). We concluded in the 1960's that there was no cause for immediate alarm concerning the past uses of the normal and isobutyl methacrylate varnishes; the polymers had not been introduced to the museum world until about 1939. The oldest coatings in 1960 would not have been more than about 20 years old. Nonetheless, in order to develop a sound understanding of the properties of acrylic resins that the conservator of museum objects might wish to employ with confidence in the future, a thorough investigation of the problem was required.

Effect of Film Thickness, Oxygen. In the range of 25- to 50-microns film thickness, the percentage of insoluble matter formed during initial stages of crosslinking decreased with film thickness. We have further shown that the crosslinking induced in these polymers by near-ultraviolet radiation is retarded when the films are exposed in an inert atmosphere and is enhanced when the polymer contains oxygen that may have been inadvertantly introduced into the chain during polymerization (5).

## Loss in Weight

During the exposure tests, we found, following an induction period in which little change was observed, that the films tended to lose weight at a rate that was linear with time. This behavior was not extensively investigated, but we believe that the induction time after which the marked rise in insoluble matter occurred corresponded closely with the time after which the distinct loss in weight began. The observation of weight loss may be of practical interest in the selection of photochemically stable acrylic coatings. A number of years ago, Berg, Jarosz and Salanthe(22) suggested that coatings that lost the minimum of weight during exposure tended to be the most durable. In a qualitative way, this was borne out in our results in the carbon-arc Fade-ometer® (Corex D filter) in which the rate of weight loss, following the induction period, was observed to be as follows: commercial normal and isobutyl methacrylate polymers, 1.0 to 2.0% loss in weight per 100 hours; poly(n-propyl methacrylate), 0.6%; Acryloid® B-72 (copolymer of ethyl methacrylate and methyl acrylate), 0.2%. The ranking is the same as the tendency-to-crosslink reported in Table I.

## Ratio of Chain Breaking to Crosslinking

_Measurement of Ratio._ Early in our investigation we encountered a commercial polymer that, although it eventually formed about 50% insoluble matter in the Fade-ometer® , did not rapidly rise to a state of more than 90% insolubility as did most of the other polymers then under investigation. The significance of this result was not fully appreciated at first; we realized later that we were observing a case of simultaneous chain breaking and crosslinking. Since then, we(4,11,23), as others before us(24), have found that the equations relating loss of solubility with exposure time that were developed by Charlesby and Pinner(25,26) for analyzing the effects of high-energy radiation also apply in the case of the photoactivated changes under study here. We have recommended to colleagues that the decrease in soluble matter during accelerated-photochemical aging tests be evaluated on a Charlesby log-log plot as seen in Figures 3 and 4. For convenience, we frequently place a scale on these graphs at ten times the gel dose to indicate the approximate ratio of chain breaks to crosslinks (Figure 3). The location of the scale divisions at ten times the gel dose are as follows: the ratio of breaks to links, $\beta/\alpha$, is 1.0 at a soluble fraction of 0.495; it is 0.7 at 0.328, 0.6 at 0.275, 0.5 at 0.225, 0.4 at 0.78, 0.2 at 0.094, and 0.1 at 0.029.

Charlesby's method of plotting $s + \sqrt{s}$ against $1/\delta$ perhaps provides a more objective analysis of the data:

$$s + \sqrt{s} = \beta/\alpha + 2/\delta$$

where s = soluble fraction, $\beta/\alpha$ the ratio of degradation to cross-

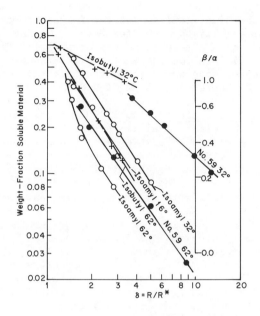

*Figure 3.    Influence of temperature on the rate of cross-linking of three methacry-late polymers plotted in terms of time relative to the gel dose R\*.  Insolubility at R/R\* = 10 used as a measure of β/α, the ratio of chain breaks to cross-links formed.  For composition of polymer 59, see Table I.*

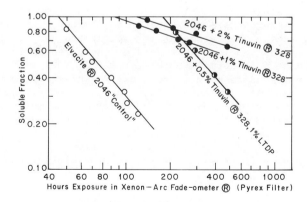

*Figure 4.    Loss of solubility of Du Pont Elvacite 2046, a 1:1 copolymer of n-butyl and isobutyl methacrylate, with and without UV absorber and antioxidant additives*

linking for the case of an initially random distribution of molecular weights, and δ is R/R*, in which R is the accumulated exposure
time and R* is the time required for the initial formation of insoluble gel, essentially the "gel dose" (25). We have used graphs
of this equation as an alternate method of plotting the data and
obtaining β/α (1,4). The locations of the reference-scale divisions
at ten times the gel dose were calculated from this equation, using
1/δ = 0.1 and the theoretical slope of 2.0.

Influence of Temperature. Panels in the carbon-arc Fade-ome-
ter® reach 62°C; the black-panel temperature in our present 600WCR
xenon-arc Fade-ometer® is 59°C. To learn how seriously temperature
affects crosslinking, we built a water-cooled surface upon which to
mount our test panels of coated aluminum foil during exposure in
the carbon-arc equipment. The results of studies at various temperatures are shown in Figure 3. Here we see that, at the usual
high temperature of panels in the carbon-arc Fade-ometer®, 62°C,
copolymer No. 59 (based on 1:3 combination of p-methylcyclohexyl
and isoamyl methacrylate) and the isobutyl polymer both tend to
crosslink almost exclusively; that is, the ratio of chain breaking
to crosslinking, β/α, ≈ 0. At a lower temperature, 32°C, considerable chain breaking occurs; the ratio of links broken to cross-
links formed becomes about 0.3 in one case, 0.55 in another. Perhaps much of this behavior can be explained by considering whether
the polymers are above or below their second-order transition temperature during irradiation, but we have not pursued this line of
investigation. As seen in Figure 3, an isoamyl methacrylate polymer based on a mixture of 2-methyl and 3-methyl esters, with its
highly labile tertiary hydrogens, crosslinked almost exclusively
even at 16°C, even though the polymer should have been below its
Tg (about 26°C) at this temperature.
    These findings clearly illustrate how the results of so-called
"accelerated-aging" tests can be affected by the high temperatures
in ordinary xenon- and carbon-arc equipment and can, therefore,
lead to erroneous conclusions regarding the photochemical behavior
of materials at near-normal temperatures (23).
    Grassie's studies on the decomposition of acrylate and metha-
crylate polymers at high temperatures may be of interest in suggesting mechanisms by which some of the alkyl groups may thermally de-
compose(27), although his investigations were conducted primarily
under vacuum or under inert gases.

Petroleum-Soluble Polymers Resistant to Crosslinking Under Near
Ultraviolet

    In view of the above findings, what may we choose for use as
a picture varnish that will have little or no tendency to crosslink? Polymers of the perfluoroacrylic esters exhibit no such tendency but require solvents of questionable appropriateness (5).
Moreover, if the coatings are used indiscriminately in making re-

pairs to the structure of a painting, future problems regarding
adhesion and wetting may be introduced. Poly(n-propyl methacrylate) is also highly resistant to crosslinking (Table I) and might
be considered, although we do not believe that a suitable commercial polymer is available.

The most stable resin for many of our purposes has proven to
be a copolymer of ethyl methacrylate and methyl acrylate. This
comes as little surprise; the Rohm and Haas Company has for years
sold a durable resin based on these two monomers, Acryloid® B-72
(6,28). We have also prepared polymers of similar physical properties based on methyl methacrylate and ethyl acrylate and have found
that their behavior is practically the same - the methyl and ethyl
groups apparently do not become seriously involved in crosslinking.
As reported elsewhere(23), rather than crosslink, Acryloid® B-72
tends to chain break under visible and near-ultraviolet radiation,
although at a very slow rate. Polyvinylacetate is another polymer
used in the care of museum objects that tends more to chain break
than crosslink under these conditions(23), but it is not our purpose to discuss its properties at this time.

One of the recognized challenges in creating a durable polymer
is to create one in which there is a proper balance between crosslinking and chain breaking tendencies. This objective was described, pursued and demonstrated in the work of Maxim and Kuist (24,
29). These authors showed that a distinct maximum in durability was
frequently obtained at an intermediate composition when monomers
were copolymerized. Much the same behavior was reported by Graham,
Crowne and MacAlpine in the evaluation of copolymers and terpoly-
mers of thermosetting acrylics (30). Maxim and Kuist's studies,
which involved acrylate rather than methacrylate polymers, confirm
our findings that crosslinking tends to increase with the length
of the alkyl side chain.

## Inhibition of Crosslinking

Our initial objectives in undertaking these investigations
were to seek an understanding of the causes of the photochemically
induced crosslinking and to pinpoint at least a few thermoplastic
polymers - originally soluble in hydrocarbons no more polar than
toluene - that had little tendency to undergo loss of solubility
under ordinary museum conditions. Having achieved these objectives, we returned our attention to the matter of inhibiting crosslinking, a practical possibility that had been accomplished through
the addition of 2,4-dihydroxybenzophenone, as described in our 1957
publication (18). We have since found that adding LTDP (dilauryl-
thiodipropionate) and Ciba-Geigy's Tinuvin® 328 ultraviolet absorber - each at 1% concentration relative to the weight of the resin-
increases the induction time 10- to 15-fold for a 1:1 copolymer of
normal and isobutyl methacrylate, Elvacite® 2046, during xenon-arc
exposure (Pyrex-glass filter) (5). Ciba-Geigy's Tinuvin®770 alone
at 1% concentration in this same resin prolongs the induction time

in the xenon-arc Fade-ometer® about 5-fold.  When plotted as we see
in Figure 4, data on loss of solubility will readily reveal which
combination of ultraviolet absorber and inhibitor affects primarily the induction time, which affects the ratio of chain breaking
to crosslinking, and which may affect both (5).

## Summary of Major Findings

The tendency to crosslink under near ultraviolet and visible
radiation has been demonstrated in polymers prepared from the n-
propyl, n-butyl and isobutyl, n-amyl and isoamyl methacrylate esters, particularly in those with alkyl groups that contain tertiary hydrogens.  The ratio of chain breaking to crosslinking has
been shown to vary with temperature, the higher temperature favoring an increase in the tendency to crosslink.  A number of toluene-
soluble polymers have been found that exhibit little tendency to
crosslink under the near ultraviolet.  Moreover, for those methacrylate polymers that do tend to crosslink, inhibitor systems can
be used to delay the onset of insolubility as much as 15-fold.  We
have determined that the reciprocity law (intensity of illumination times induction time) essentially holds true over a 30-fold
range of intensity.  Further, the rate of formation of insoluble
matter has been shown to increase logarithmically with the decrease
in the lowest wavelength of irradiance in the range of 400 to 300
nm.  Crosslinking of these polymers can take place under the normal
conditions of daylight or fluorescent-lamp illumination encountered
in a museum, although at an extremely low rate.

## Importance with Respect to Long-Term Usage

These studies, sponsored for over two decades by the National
Gallery of Art, have established some general principles that can
guide conservators both in the selection of durable protective
coatings and in the application, maintenance, and repair of such
coatings.  Based upon tests made under the relatively mild conditions of exposure to "daylight" fluorescent lamps and also to natural illumination on a gallery wall, we can now reasonably predict
that - on an art gallery wall experiencing about 110,000 footcandle
hours of diffuse daylight through window glass annually - polymers
such as Acryloid® B-72 and polyvinylacetate should remain color-
less, and soluble in the solvents in which they were originally
soluble, for more than 200 years (31).  Truly, these are first-class
materials to have been placed at the artists' and conservators'
command.

## Acknowledgements

We are grateful to the organizers of this symposium for the
opportunity to enlarge upon our reports of these investigations
that have appeared in a wide variety of publications in the past,

publications that have been directed more towards the museum conservator and conservation scientist than to the polymer chemist.
Many of the methacrylate polymers used at the beginning of
these investigations were prepared by Stuart Raynolds. Richard A.
Tauson carried out the initial studies on the effect of temperature. The authors wish particularly to thank Dr. John Walker, Di-
rector Emeritus of the National Gallery of Art, for his great per-
sonal interest, support, and encouragement in this research. The
work of the Center is principally made possible through the gener-
osity of the Andrew W. Mellon Foundation.

## Literature Cited

1. Feller, R. L., Stolow, N. and Jones, E. B., "On Picture Varnishes and Their Solvents", Press of the Times, Oberlin, Ohio,
   1959; revised edition of Press of Case-Western Reserve, Cleveland, 1971; both currently out of print.
2. Raynolds, S., "The Dependence of Physical Properties on the
   Constitution of Alkyl Polymethacrylate-ester Resins", Thesis,
   Master of Science, University of Pittsburgh, 1954.
3. Feller, R. L., "Identification and Analysis of Resins and
   Spirit Varnishes", in "Application of Science in Examination
   of Works of Art", Museum of Fine Arts, Boston, 1959, pp. 51-76.
4. Feller, R. L. and Bailie, C. W., "Studies of the Effect of
   Light on Protective Coatings Using Aluminum Foil as a Support:
   Determination of Ratio of Chain Breaking to Cross-linking",
   Bulletin of the American Group-IIC, 1966, 6, No. 1, 10-12.
5. Feller, R. L., "Problems in the Investigation of Picture Varnishes", in "Conservation and Restoration of Pictorial Art",
   Eds. Brommelle, N. and Smith, P., Butterworths, 1976, pp. 137-
   144.
6. Feller, R. L., "New Solvent Type Varnishes", in "Recent Advances in Conservation", Butterworths, London, 1963, pp. 171-175.
7. Miller, C. D., "Kinetics and Mechanism of Alkyd Photooxidation", Ind. Eng. Chem., 1958, 50, 125-128.
8. Harrison, L. S., "An Investigation of the Damage Hazard in
   Spectral Energy", Illum. Eng. (NY), 1954, 49, 253-257.
9. Bateman, L., "Photolysis of Rubber", J. Polym. Sci., 1947, 2,
   1-9.
10. Martin, K. G. and Tilley, R. I., "Influence of Radiation Wave-
    length on Photooxidation of Unstabilized PVC", Br. Polym. J.,
    1971, 3, 36-40.
11. Feller, R. L., "The Deterioration of Organic Substances and
    the Analysis of Paints and Varnishes", in "Preservation and
    Conservation, Principles and Practices", Smithsonian Press,
    Washington, D.C., 1976, pp. 287-299.
12. Feller, R. L., "Control of Deteriorating Effects of Light on
    Museum Objects", Museum, 1964, 17, 57-98.
13. Feller, R. L., "Speeding Up Photochemical Deterioration",
    Bulletin royal du Patrimoine artistique, 1975, 15, 135-150.

14. Morimoto, K. and Suzuki, S., "Ultraviolet Irradiation of
    Poly(alkyl Acrylates) and Poly(alkyl Methacrylates)", J. Appl.
    Polym. Sci., 1972, 16, 2947-2961.
15. Grassie, N. and MacCallum, J. R., "Thermal and Photochemical
    Degradation of Poly(n-butyl Methacrylate)", J. Polym. Sci.,
    1964, 2, Part 2A, 983-1000.
16. Grassie, N., "Photodegradation of Methacrylate/Acrylate Co-
    polymers", Pure and Appl. Chem., 1973, 34, 247-257.
17. Oster, G., "Photochemical Crosslinking of Non-Aqueous Poly-
    mers by Near Ultraviolet Light", J. Polym. Sci., 1964, 2,
    Part B, Polymer Letters, 1181-1182.
18. Feller, R. L., "Cross-Linking of Methacrylate Polymers by Ul-
    traviolet Radiation", Preprints of papers presented at the
    New York Meeting, Division of Paint, Plastics and Printing Ink
    Chemistry, American Chemical Society, Sept., 1957, 17, No. 2,
    465-470.
19. Barton, J., "Peroxide Crosslinking of Poly(n-alkyl Methacry-
    lates)", J. Polym. Sci., 1968, 6, Part 1A, 1315-1323.
20. Gordon, G. Ya., "Stabilization of Synthetic High Polymers",
    Israel Program for Scientific Translations, 1964, p. 45.
21. Chien, J. C. W. and Boss, C. R., "Polymer Reactions. V. Ki-
    netics of Autoxidation of Polypropylene", J. Polym. Sci.,
    1967, 5, Part 1A, 3091-3101.
22. Berg, C. J., Jarosz, and Salanthe, G. F., "Performance of
    Polymers in Pigmented Coatings, J. Paint Technol., 1967, 39,
    436-453.
23. Feller, R. L., "Stages in the Deterioration of Organic Materi-
    als", in Williams, J. C., "Preservation of Paper and Textiles
    of Historic and Artistic Value", Advances in Chemistry Series
    No. 164, American Chemical Society, 1977, pp. 314-335.
24. Maxim, L. D. and Kuist, C. H., "The Light Stability of Vinyl
    Polymers and the Effect of Pigmentation", Off. Dig., Fed. of
    Soc. for Paint Technology, 1964, 36, 723-744.
25. Charlesby, A. and Pinner, S. H., "Analysis of the Solubility
    Behavior of Irradiated Polyethylene and Other Polymers", Proc.
    R. Soc., 1959, A249, 367-386.
26. Charlesby, A., "Atomic Radiation and Polymers", Pergamon
    Press, London, 1960, pp. 143, 173.
27. Grassie, N., "Recent Work on the Thermal Degradation of Acry-
    late and Methacrylate Homopolymers and Copolymers", Pure and
    Appl. Chem., 1972, 30, 119-134.
28. DeWitte, E., Goessens-Landrie, M., Goethals, E. J. and Simons,
    R., "The Structure of 'Old' and 'New' Paraloid B-72", Inter-
    national Council of Museums (Paris) Committee for Conserva-
    tion, Zagreb Meeting, 1978, Paper 78/16/3.
29. Kuist, C. H. and Maxim, L. D., "The Ultraviolet Degradation
    of Scissioning Copolymers", Polymer, 1965, 6, 523-530.
30. Graham, N. B., Crowne, F. R. and MacAlpine, D. E., "A Maximum
    Durability Prediction Scheme for Thermosetting Acrylics", Off.
    Dig., Fed. of Soc. for Paint Technology, 1965, 37, 1228-1250.
31. Feller, R. L. and Curran, M., "Changes in Solubility and Re-
    movability of Varnish with Age", Bulletin American Institute
    for Conservation, 1975, 15, No. 2, 17-48.

RECEIVED October 28, 1980.

# Photodegradation of Polyvinyl Chloride

## A Survey of Recent Studies

W. H. STARNES, JR.

Bell Laboratories, Murray Hill, NJ 07974

Although poly(vinyl chloride) (PVC) is one of the most important commercial polymers, its outdoor use has been restricted by its photochemical instability. The reasons for this instability are incompletely understood, but some progress has been made recently on this problem, and the present paper attempts to summarize the current status of fundamental knowledge in this field. This survey is not intended to be comprehensive; it is concerned primarily with work published since the early 1970's and with basic chemical principles rather than technological developments. The photodegradation of PVC has been discussed in other recent reviews (1,2,3,4).

General Considerations

Photolysis of PVC in the presence of oxygen causes oxidation of the polymer. However, under most (perhaps all) conditions, in both the presence and absence of oxygen, the photodegradation is complicated by scissions of carbon-chlorine bonds. Such scissions may lead to the formation of conjugated polyene sequences via sequential dehydrochlorination (Equation 1). The polyenes

$$-(CH_2CHCl)_n- \xrightarrow{h\nu} -(CH=CH)_n- + \underline{n}HCl \qquad (1)$$

and the initial oxidation products may undergo photooxidation, in turn, or they may be destroyed by photochemical processes of the nonoxidative variety. Thermal reactions can occur also when temperatures are sufficiently high. Thus the photodegradation of PVC produces extremely complex chemical systems whose compositions are difficult to determine and whose behavior is hard to predict. It is, therefore, hardly surprising to find that most of the basic problems in this field have not been solved.

Recent fundamental studies in this area have been concerned with aspects such as the nature of the initiating chromophores, the chemistry of the initiation steps, the extent to which the

0097-6156/81/0151-0197$05.00/0

chromophores and other reactive groups are formed during polymerization and subsequent processing, the products and reaction mechanism of the overall degradation process, and, of course, the
effects of experimental variables (wavelength and intensity of the
incident light, oxygen pressure, temperature, time, sample purity,
and sample thickness). Some of these investigations have been
carried out with light at wavelengths below the terrestrial solar
range (i.e., at wavelengths < ∿290 nm) (5), and it should be kept
in mind that results obtained under circumstances such as these
may not be strictly applicable to natural weathering situations.

## Initiation

The ordinary monomer units of PVC are not expected to absorb any terrestrial solar radiation (1,2,3,4). Thus, under the
usual ambient conditions, photodegradation of the polymer must be
initiated by chromophoric impurities. These impurities may simply be structural defects in the PVC itself, or they may be extraneous substances that have been incorporated into the polymer.
Several of these potential photosensitizers are discussed in the
following sections.

Carbonyl Groups. Such structures could be introduced by air
oxidation during polymerization or subsequent processing of the
polymer. There is, in fact, some experimental evidence for their
formation during polymerization via the following sequence of
steps (6): (1) copolymerization of vinyl chloride with adventi-
tious oxygen; (2) decomposition of the resulting polyperoxide to
form HCl, $CH_2O$, and CO; (3) copolymerization of CO with vinyl
chloride.

Free-radical copolymers of vinyl chloride with carbon monox-
ide have been suggested to contain pendent COCl groups (as in $\underset{\sim}{1}$)

$$-CH_2CH- \qquad\qquad -CH_2-CHCl-CO-CH_2-CHCl-$$
$$\quad\;\; |$$
$$\quad\; COCl$$
$$\quad\;\; \underset{\sim}{1} \qquad\qquad\qquad\qquad\qquad \underset{\sim}{2}$$

rather than backbone carbonyls (7). However, other work has indicated that the latter type of structure (2) may be correct (8,
9). In any event, studies with authentic poly(vinyl chloride-
co-carbon monoxide) polymers have shown that the carbonyl groups
do, indeed, accelerate the photodehydrochlorination and photooxidation of these materials (9,10,11). Moreover, benzophenone
is known to act as a photosensitizer for the dehydrochlorination
of PVC (12), and the dehydrochlorination of several simple alkyl
chlorides has been found to be photosensitized by benzophenone
(13,14,15) and a number of alkyl aryl ketones (13,14). Self-
sensitized photodehydrochlorination has been observed with
4-chloroalkyl phenyl ketones (16,17,18). Acetone (13,19,20,21)

and several other aliphatic ketones (13,19) have also been shown
to photosensitize the elimination of HCl from PVC (19) and simple
alkyl chlorides (13,19,20,21), although the sensitization efficiency of such ketones declines precipitously when their alkyl
groups are large (13).    Another effective ketone sensitizer for
the photodehydrochlorination of the polymer is hexachloroacetone
(22).

Several mechanisms have been postulated in order to account
for ketone-sensitized photodehydrochlorination.    Benzophenone
and acetophenone have been suggested to act as singlet sensitizers via a collisional deactivation process (13).    An alternative mechanism proposed for benzophenone involves abstraction of
a methylene hydrogen from PVC by the triplet ketone (Equation 2),
followed by β scission of a

$$^3Ph_2CO^* \ + \ -(CH_2CHCl)_\underline{n}- \ \longrightarrow \ Ph_2\overset{\bullet}{C}OH \ + \ -\overset{\bullet}{C}HCHCl-(CH_2CHCl)_{\underline{n}-1}- \quad (2)$$

$$\underset{\thicksim}{\overset{3}{}}$$

$$-\overset{\bullet}{C}HCHCl-(CH_2CHCl)_{\underline{n}-1}- \ \longrightarrow \ Cl\cdot \ + \ -CH=CH-(CH_2CHCl)_{\underline{n}-1}- \quad (3)$$

chlorine atom from the resulting carbon radical (Equation 3) (12).
This mechanism seems consistent with the appearance of an absorption band at 340 nm that can be assigned to an adduct of benzo-
phenone and radical $\underset{\thicksim}{3}$ (12), and with the ability of naphthalene
(a triplet quencher) to retard both the rate of appearance of
this absorption and the rate of disappearance of benzophenone
(23).    Arguments have been given for the operation of a similar
mechanism involving intramolecular hydrogen abstraction in the
case of the 4-chloroalkyl phenyl ketones (16,17,18).    On the
other hand, quenching studies with tert-butyl chloride have indicated that the hydrogen-abstraction mechanism is actually quite
unlikely for dehydrochlorinations that are photosensitized by
benzophenone and alkyl phenyl ketones in an intermolecular manner
(14).    An alternative scheme that can be proposed for such reactions invokes an intermediate (triplet ketone)-substrate exciplex whose decomposition (Equation 4) produces a vibrationally

$$^3Ph_2CO^* \ + \ RCl \ \rightleftarrows \ ^3[exciplex]^* \longrightarrow \ Ph_2CO \ + \ RCl^*_{vib} \quad (4)$$

$$RCl^*_{vib} \longrightarrow \ R\cdot \ + \ Cl\cdot \ \longrightarrow \ HCl \ + \ alkene \quad (5)$$

(R = alkyl)

excited alkyl chloride that experiences dehydrochlorination,
either in a concerted manner (14) or via a route involving the
disproportionation of free-radical intermediates (e.g., Equation
5; solvent radicals could also become involved in this process).
The exciplex was originally suggested to be stabilized by a
charge-transfer interaction in which the alkyl chloride is the

electron donor (14), but later workers have argued that the chloride moiety is the electron acceptor instead (15). At any rate,
it would now seem that the same type of mechanism could apply to
benzophenone and PVC, and that the disappearance of benzophenone
and the appearance of the 340-nm absorption which have been noted
in that system (12,23) might actually signify the occurrence of
Reaction 6 or Reaction 7 [cf. (12) and references cited therein].

$$Ph_2CO \quad + \quad R\cdot \quad \longrightarrow \quad Ph_2\overset{\cdot}{C}OR \tag{6}$$

$$Ph_2CO \quad + \quad R\cdot \quad \longrightarrow \quad \xrightarrow{+(H\cdot)} \quad PhC\overset{OH}{\underset{R}{=}}\langle\underset{=}{=}\rangle\overset{H}{} \tag{7}$$

(R = a polymeric carbon radical)

The acetone-sensitized photodehydrochlorination of 1,4-
dichlorobutane is not suppressed by triplet quenchers (20), but
the fluorescence of the sensitizer is quenched by the alkyl
chloride (13). These observations imply the operation of a mechanism involving collisional deactivation, by the substrate, of
the acetone excited singlet state (13,21). This type of mechanism has received strong support from another study in which the
fluorescence of acetone and 2-butanone was found to be quenched
by several alkyl and benzyl chlorides (24). The detailed mechanism for alkanone sensitization proposed on the basis of the latter work invokes a charge-transfer (singlet ketone)-substrate
exciplex (24) and is similar to one of the mechanisms that has
been suggested (15) for sensitization by ketone triplets (cf.
Equations 4 and 5).

Another mechanism for alkanone-sensitized photodehydrochlorination comprises Norrish type I scission of the ketone,
followed by ground-state reactions of radicals (19). However,
the evidence for such a mechanism is based on experiments that
were carried out in the vapor phase (19). Initiation of the
photodegradation of PVC by hexachloroacetone has been suggested
to involve the abstraction of hydrogen from the polymer by radicals resulting from the photolysis of the ketone's carbon-
chlorine bonds (22).

Several recent workers have considered the possibility that
the photooxidation of PVC involves oxidation by singlet oxygen
which results from the $^3O_2$ quenching of triplet carbonyl impurities (2,9,25). This type of mechanism has neither been
established nor disproven, although, as expected, $^1O_2$ appears to
be essentially unreactive toward undegraded PVC (26). Other recent observations that may be pertinent to the carbonyl-sensi-
tized photodehydrochlorination of the polymer are the failure of
β-chloropropiophenone (16) and γ-chlorobutyrophenone (16,17) to
undergo photodehydrochlorination [the latter ketone experiences
Norrish type II scission instead (17)] and the occurrence of type
II scission (with no concomitant type I cleavage) upon irradiation of 5-chloro-2-hexanone (27).

As noted above, there is experimental evidence to indicate
that the carbonyl groups of (vinyl chloride)-(carbon monoxide)
copolymers are effective sensitizers for the photodegradation of
these materials (9,10,11). A reasonable sensitization mechanism
can be formulated for this system on the basis of the information
now on hand.

Photodehydrochlorination of poly(vinyl chloride-co-carbon
monoxide) is not accompanied by significant changes in polymer
molecular weight when it is carried out under air with a high-
pressure mercury lamp (11,28). It also occurs under nitrogen at
wavelengths > 294 nm with no appreciable alterations in the position and intensity of the original IR carbonyl absorption (9).
Thus the photoinitiation occurring in this system does not seem
to require Norrish scissions or any other permanent structural
changes in the immediate vicinity of the sensitizing carbonyl
groups. Moreover, the initial stage of the photodehydrochlorination is strongly inhibited by $^3O_2$ (9,25), a result which suggests
[by analogy with other work (20)] that alkanone excited singlets
are not involved in the sensitization process. These findings
would seem to be in keeping with an initiation mechanism (Equations 8 and 9) that is

$$
\begin{aligned}
&\overset{3}{\underset{\|}{O}}{}^* \\
&-\overset{\|}{C}- \;+\; -CH_2CHCl- \;\rightleftharpoons\; {}^3[\text{exciplex}]^* \longrightarrow -\overset{\overset{O}{\|}}{C}- \;+\; -(CH_2CHCl)^*_{vib}- \quad (8)
\end{aligned}
$$

$$
-(CH_2CHCl)^*_{vib}- \;\longrightarrow\; -CH=CH- \;+\; HCl \quad (9)
$$

operable at ambient solar wavelengths and involves exciplex formation from a carbonyl triplet and a chloromethylene-containing
portion of the polymer, followed by a dissociation process that
regenerates ground-state carbonyl and produces a vibrationally
excited polymer segment that eliminates HCl in a stepwise or
concerted manner. Other possible variations of the mechanism
would involve the direct occurrence of dehydrochlorination from
the exciplex intermediate, or dehydrochlorination following dissociation of the exciplex into a cation-anion radical pair. This
scheme is, in fact, directly analogous to the aryl ketone mechanism of Equations 4 and 5 (14,15).

In the case of ordinary commercial PVC, the importance of
carbonyl photosensitization is not entirely clear. Its occurrence is consistent with the reported ability of $^3O_2$ to retard
the initial photodehydrochlorination of PVC itself (9,29).
Nevertheless, this evidence is not conclusive, and other work,
discussed in the following section, has suggested that another
sensitization mechanism predominates with PVC under some conditions, at least.

Alkene Linkages. Using (250-350)-nm irradiation (30,31)
or the unfiltered light from a high-pressure mercury lamp (32),
Balandier and Decker have measured quantum yields under nitrogen

and oxygen atmospheres for the dehydrochlorination, chain scission, and cross-linking of PVC in solution (30,31) and in films
(31,32). The values obtained were found to be independent of
the extent of reaction, despite the increasing absorption of
light by the polyene structures that were formed (30,31,32).
Thus the constancy of the quantum yields was taken as evidence
for the initiation of photodegradation by unsaturated structures
in the original polymer (30,31,32). Since the quantum efficiencies were not affected significantly by changes in light intensity or the polymer concentration (30), the photodegradation
appears to have been primarily a non-chain intramolecular process.
Polyene triplet states are unlikely to have been involved in the
mechanism, owing to the low triplet energies (33) and very low
intersystem crossing efficiencies (34) expected for chromophores
of this general type. Moreover, the photodehydrochlorination
showed no evidence of inhibition by triplet molecular oxygen (30,
31). All of these observations seem consistent with the simple
process of Reaction 10 (which might involve the disproportiona-

$$-{}^{1}(CH=CH)_{\underline{n}}^{*}-CHClCH_2- \longrightarrow -(CH=CH)_{\underline{n+1}}- + HCl \qquad (10)$$

tion of radicals resulting from C–Cl homolysis). Other arguments
in support of such a mechanism have been presented by Reinisch
et al. (35), and here it is also of interest to note that the
unsensitized photolysis of liquid allyl chloride appears to involve C–Cl homolysis as the primary photochemical process (36).

Is alkene sensitization important for the PVC polyenes that
absorb at ambient solar wavelengths? The available information
does not provide an unambiguous answer to this question. Decker
found that the average length of the polyenes increased with increasing time of irradiation and caused an enhanced absorption
of the incident light having wavelengths greater than 366 nm
(32). This observation, together with the time independence of
the dehydrochlorination quantum efficiency, could be taken as evidence for photosensitization by polyenes absorbing in the ambient
solar range (32). Furthermore, a comparable dehydrochlorination
quantum yield was obtained from the 514.5-nm laser photolysis of
a PVC film that had been degraded previously by UV irradiation
under nitrogen (32). On the other hand, Gibb and MacCallum observed that the polyenes formed from PVC upon irradiation under
nitrogen at 240–560 nm did not effectively sensitize further
photodegradation during subsequent irradiation under nitrogen
with higher-wavelength light (37). Redistribution of the polyene
sequence lengths was observed instead (37), and it was concluded
from these findings that the principal initiating chromophores
are conjugated dienes and trienes when > 240-nm irradiation is
employed (37). Other workers have reported that the rate of the
photodehydrochlorination of PVC is so low at wavelengths >310 nm
that long conjugated polyenes in the polymer ($\underline{n}$ > 3 in Equation 1)

can be analyzed for quantitatively via their selective photooxidation using (310–450)-nm light (38).

The reason for these divergent results is, at present, unclear. However, in view of the difference noted above with regard to inhibition by triplet oxygen, it seems that the photosensitization mechanism for the systems of Balandier and Decker (30,31,32) must have differed from that which operated in the systems of Braun et al. (9,25). For the photodegradations performed with simulated terrestrial insolation, it can be argued that carbonyl groups were the initial active sensitizers, and that long polyenes assumed the predominant role of sensitization in the latter stages of the process. Further measurements of quantum yields at ambient solar wavelengths might help to settle this point, especially if the measurements were performed on polymers containing different amounts of polyene structures. In this connection, it should be noted that polymers containing long polyenes are extremely susceptible to thermal dehydrochlorination (39,40) (apparently at temperatures as low as 30°C!) (40), and that this process may complicate quantum yield measurements in some situations, particularly if a focused laser beam is used as the source of the incident light.

Alkene sensitization of the photodegradation of PVC seems to be supported by other observations that have been reported in the recent literature. One of these is the increased rates of nonoxidative photodegradation that have been found for polymers which were subjected to a preliminary nonoxidative thermal treatment (41,42). Moreover, the benzophenone–photosensitized dehydrochlorination of PVC has been shown to undergo an autoacceleration which can be attributed to a supplementary photosensitization by the conjugated polyene product (12). Also of interest in this regard is the finding that the production of free radicals during the UV irradiation of PVC under vacuum apparently can be sensitized by polyenes that have been created during a preliminary photolysis (43,44).

A correlation has been reported between the photooxidizability of several PVC samples and the number of long-chain ends in these polymers (45). The mechanistic significance of this result is rather uncertain, as the authors (45) do not state how the number of long-chain ends was varied. Nevertheless, it is perhaps worth noting that, in the absence of added chain-transfer agents, a major fraction of the long-chain ends should consist of allylic chloride groups (46,47).

Catalytic hydrogenation of PVC causes a significant reduction in the rate of the polymer's subsequent photodegradation (32). Although this result is consistent with the occurrence of alkene sensitization (32), it can also be attributed, per se, to the removal of other possible sensitizers such as carbonyl groups and peroxide linkages.

On the basis of spectroscopic observations, Verdu et al. (48) have argued that polyene sequences are the principal

sensitizers for the photooxidation of (vinyl chloride)-(carbon
monoxide) and (vinyl chloride)-oxygen copolymers at 300-450 nm.
For the case of the poly(vinyl chloride-co-carbon monoxide)s,
this conclusion is difficult to reconcile with the different
initial kinetic effects of $^3O_2$ that were found for the systems
of Braun et al. (9,25) as compared to those of Balandier and
Decker (30,31).

Alkene linkages that are formed during processing seem to
play an important role in the photooxidation of PVC (49,50,51),
and some consideration has been given recently to the possibility
of a degradation mechanism involving energy transfer from excited carbonyl groups to alkene linkages in the polymer (9).
On the other hand, carbonyl quenching by polyenyl radicals in
PVC has been invoked in order to account for autoinhibition during photooxidative degradation (52). Some workers have mentioned
the possibility of singlet oxygen formation in PVC via the $^3O_2$
quenching of excited polyenes (25,29) or cyclohexadienes (53).

Peroxides. Direct or sensitized photolysis of an oxygen-
oxygen linkage would produce alkoxy radicals that might initiate
the degradation of PVC. Indeed, the peroxidic gel fraction of
mildly processed polymer has been found to be extremely susceptible to subsequent photooxidation (49). The hydroperoxide
concentration of such a gel can be correlated with the initial
photooxidation rate (49), but the olefinic unsaturation produced
concurrently during processing also appears to contribute to the
photooxidative instability of the polymer (50,51). Enhanced
rates of photodegradation have been reported also for PVC prepared in the presence of oxygen (9,10,48) and, at λ >320 nm, for
PVC containing peroxide that was introduced by ozonization (52).
In these cases, both carbonyl groups and peroxides may have
contributed to the photoactivation effect [however, see (48)].

Several carbonyl-containing peroxide additives have been
shown to increase the initial rate of the nonoxidative photodehydrochlorination of PVC (54). In studies with polymeric ketones unrelated structurally to PVC, the excited singlet and
triplet states of the carbonyl groups in these polymers were
found to sensitize 0-0 homolysis at rates approaching diffusion
control (55). Similar reactions may well occur in oxidized
vinyl chloride polymers.

Solvents, Additives, and Extraneous Impurities. The rate of
PVC photodegradation under air at wavelengths >250 nm is said to
be increased by small amounts of residual tetrahydrofuran (THF)
or dichloroethane (56). On the other hand, residual THF has
been reported not to enhance the dehydrochlorination of the
polymer during irradiation under nitrogen at λ >240 nm (41).
Nevertheless, under the conditions of the latter study, THF
was found to increase the relative concentrations of the shorter
polyene products (41). This effect was attributed to a facile

rotation around the C-C bonds of the longer polyene sequences, owing to the ability of THF to function as a plasticizer (41).

Tetrahydrofuran has been reported to exhibit an absorption maximum at 280 nm (52,56), but several workers have shown that this band is not produced by the purified solvent (30,41,57). Oxidation products from THF have been invoked in order to account for the appearance of the 280-nm band in PVC films that are solvent-cast from THF in air (57, 58). However, in some reported cases (56,59), this band was undoubtedly produced, at least in part, by a phenolic antioxidant (2,6-di-tert-butyl-p-cresol)(59) in the solvent. Since certain p-alkylphenols have now been shown to be powerful photosensitizers for the dehydrochlorination of PVC (60), it is clear that antioxidant photosensitization might well have been responsible for some of the effects attributed previously (56) to THF alone. On the other hand, enhanced rates of photodegradation under air have also been observed for PVC films cast from purified THF (57), a result which has been ascribed to radical formation during the photooxidation of residual solvent (57,61). Rabek et al. (61) have shown that this photooxidation produces $\alpha$-HOO-THF, $\alpha$-HO-THF, and $\gamma$-butyro-lactone, and they have found that the hydroperoxide product is an effective sensitizer for the photodehydrochlorination of PVC at $\lambda$ = 254 nm (61).

Studies with model compounds have demonstrated that photo-dehydrochlorination is sensitized by p-cresol triplets via a charge-transfer exciplex intermediate in which the alkyl chloride is the electron acceptor (15). The detailed mechanism suggested for this process (15) is outlined in Equations 11 and 12.

$$^3AroH^* \; + \; RCl \; \longrightarrow \; ^3[\overset{\delta+}{AroH}\text{--}\overset{\delta-}{RCl}]^* \; \longrightarrow \; RCl\overline{\cdot} \quad\quad (11)$$

$$RCl\overline{\cdot} \; \longrightarrow \; Cl^- \; + \; R\cdot \; \longrightarrow \text{alkenes, etc.} \quad\quad (12)$$

$$(R = alkyl)$$

A number of recent studies have been concerned with the effects of commercial heat stabilizers on the photodegradation of PVC. During irradiation at room temperature under air with 253.7-nm light, several dialkyltin dicarboxylates were found to increase the rates of the photooxidation and cross-linking of the polymer (62). However, at 0°C under air in a sunshine weatherometer, photooxidation was shown to be retarded by certain dibutyltin dicarboxylates (63). The latter result was also obtained in experiments involving the use of dibutyltin maleate with irradiation in the (280-400)-nm wavelength region at 38°C under air (50,51,64,65,66).

Varying effects have been observed as well for stabilizers of the dialkyltin bis("isooctyl" thioglycolate) type. These substances accelerate PVC photooxidation (62,66) and cross-linking (62) under some conditions. Moreover, the dioctyl derivative was found to have a sensitizing effect on the photo-dehydrochlorination of PVC in experiments performed at 20°C under nitrogen with a high-pressure mercury lamp (67). However, in another study that was carried out under air with > 300-nm wavelength light (68), photooxidation of the polymer was strongly retarded by dibutyltin bis("isooctyl" thioglycolate).

Most workers agree, at any rate, that dialkyltin heat sta-bilizers can reduce the discoloration of PVC during photooxi-dation (51,62,65,66,67,69), although this conclusion is at var-iance with some results that have been reported for dibutyltin bis("isooctyl" thioglycolate) (68). A possible explanation for this discrepancy is that SnS (a black substance) was formed by photolysis of the thioglycolate stabilizer (62) at long irradia-tion times. Dialkyltin stabilizers, in general, do experience photodecomposition (62,66), and such a process has been demon-strated to occur with unusual facility for the sulfoxide formed in situ from dioctyltin bis("isooctyl" thioglycolate) (66).

Other work in this general area has shown that the photo-oxidation of PVC is retarded by zinc and cadmium stearates (63) and by $Bu_2SnCl_2$ (68), although it is accelerated by $Bu_2SnO$ and $Bu_2SnCl(SCH_2CO_2$-"isooctyl") (68). Butyltin trichloride first retards and then accelerates the photooxidation process (68). A comprehensive discussion of the photostabilization of PVC is outside the scope of the present chapter, and the reader is referred to other reviews (1,2,4) for coverage of this subject.

Water, methanol, and n-hexane do not influence the photo-oxidation of PVC (43), but the photodegradation is accelerated by ferric chloride (70,71) and certain other compounds contain-ing iron (70,71,72). Purification of the polymer might be ex-pected to enhance its photostability by removing deleterious impurities such as iron compounds that are derived from metal equipment. This type of result was obtained in one recent study (58) but not in others (30,59). In contrast, the photo-oxidative degradation of PVC should be enhanced by admixture of the polymer with materials that are unusually susceptible to photooxidation themselves. Such behavior has been observed for impact-modified PVC containing polybutadiene-based polyblends (69,73).

## Structural Defects Determined in PVC by Carbon-13 NMR

Reductive dehalogenation of PVC or PVC-$\alpha$-d with chemical reducing agents ($Bu_3SnH$, $Bu_3SnD$, $LiAlH_4$, or $LiAlD_4$), followed by C-13 NMR analysis of the reduction products, has provided much insight into the nature and concentration of the structural defects formed during vinyl chloride polymerization. Some of these defects undoubtedly influence the photostability of the polymer--in some cases, by simply acting as labile starting points for the growth of polyene sequences during processing. It therefore seems appropriate to summarize these studies here.

The principal defect identified thus far [(2-3)/(1000 C)]
is -CH$_2$-CH(CH$_2$Cl)-CHCl- (46,47,74-79). Under nonoxidative conditions, this structure probably has no appreciable effect upon
the thermal or photochemical stability of PVC. However, it
may facilitate the photooxidation of the polymer, since it contains a tertiary hydrogen atom. Other branch structures, present in lesser amounts, are -CH$_2$-CCl(CH$_2$-CH$_2$Cl)-CH$_2$- (80,81);
the long-branch array, -CH$_2$-CCl(CH$_2$-)-CH$_2$- (80,81,82); and
(probably) -CH$_2$-CCl(CH$_2$-CHCl-CH$_2$-CH$_2$Cl)-CH$_2$- (81). The tertiary
halogen atom in these groups should expedite their dehydrochlorination at typical processing temperatures (83,84). Many
of the long-chain ends appear to consist of allylic chloride
moieties (46,47,80,81) which should also be thermally labile
(83,84); the most probable structure for these groups is
-CH$_2$-CH=CH-CH$_2$Cl (46,80,81). Saturated long-chain ends are of
two types, -CHCl-CH$_2$-CHCl-CH$_2$Cl and -CH$_2$-CHCl-CH$_2$-CH$_2$Cl (46,47,
80,81,82); they appear in a mole ratio of ca. 75:25, respectively,
for polymers prepared at 100°C in bulk (80,81). Neither of the
saturated ends is likely to have a major effect upon the thermal
or photochemical stability of the polymer, although the thermal
stability of the 1,2,4-trichlorobutyl end may be somewhat less
than that of the ordinary monomer units. There is as yet no
C-13 NMR evidence for the presence of monomer head-to-head em-
placements (46), and the number of internal allylic groups seems
to be very small in commercial polymers (75,76,77,78).

## Overall Course and Reaction Mechanism of PVC Photodegradation

The latter stages of the photodegradation of PVC are not
well understood and are at least as controversial as the nature
of the photoinitiation steps. Mechanisms proposed for the
overall reaction have remained primarily speculative, owing in
part to a lack of definitive information about the structural
changes taking place. For this reason, no attempt will be made
to present a comprehensive mechanism here. What will be considered instead are some of the more significant observations
that have been made in recent years on certain aspects of this
problem.

One of these aspects is the question of whether dehydrochlorination always occurs during the photodegradation of PVC.
The evidence on this point is confusing, but it implies, in any
event, that the extent of dehydrochlorination is dependent on
sample temperature and the wavelength of the incident light. As
noted above, data exist to suggest that certain photoreactions
of the polyenes in PVC [redistribution of sequence lengths (37)
and photooxidation (25,38)] can occur without any appreciable
dehydrochlorination when long-wavelength light is used. Also,
in some investigations performed recently with a carbon-arc
weatherometer ($\lambda \geq \sim$280 nm), the polymer was found to photo-
oxidize extensively at 0°C with no detectable loss of HCl (58,
85), although dehydrochlorination did take place at higher

sample temperatures (85). Another report has indicated that de-hydrochlorination does not ensue during photooxidation at 30°C with 253.7-nm light (86). However, other workers have detected a significant amount of photodehydrochlorination during irradia-tion at 30°C under air with a high-pressure mercury lamp (11), and they have found that the rate of this process increases with temperature in the 50-90°C range (11). In fact, spectral data have suggested that polyenes are formed even at -196°C during irradiation of the polymer with 253.7-nm light (43,87).

Activation energies that have been reported for the photo-dehydrochlorination of PVC are 18 kJ mol$^{-1}$ (at -40 to +70°C) (35), 40.5 kJ mol$^{-1}$ (at -20 to 0°C) (88), 14 kJ mol$^{-1}$ (at 0 to +60°C) (89), and 8.3 kJ mol$^{-1}$ (at +20 to +90°C) (88). The last value has been stated to be independent of oxygen concentration (25), and a value of 21 kJ mol$^{-1}$ has been determined for the photooxidation of the polymer at +20 to +90°C (25). These re-sults indicate that decreases in temperature should favor photo-dehydrochlorination over photooxidation when the temperature is $\geq$ 0°C. Yet just the opposite effect was observed in the work of (85). At present this dichotomy defies a rational explanation.

There is one report of polyene formation during photooxida-tion with no concurrent evolution of HCl (52)! This observa-tion can perhaps be attributed to insensitivity of the method used for HCl detection.

Many studies relate to the effect of irradiation wavelength per se. Irradiation of virgin PVC under nitrogen with (266-370)-nm light has been reported to cause only a negligible amount of dehydrochlorination (37). However, other investigators have observed appreciable photodegradation of the polymer at wavelengths > 300 nm (10,19,30,52,59). Most workers agree, at any rate, that degradation increases with decreasing wavelength (10,19,30,37,52,56), although one report (52) has suggested that this conclusion may not apply to the separate rates of C=C and C=O formation when monochromatic light is used at wave-lengths above 300 nm. Braun and Kull (88) have recently em-phasized the important point that an apparent wavelength depend-ence may actually be an irradiation intensity dependence in some experimental situations.

When molecular oxygen is absent, the average length of the conjugated polyenes formed during UV irradiation of PVC has been found to increase with increasing temperature (88,90) and in-creasing syndiotacticity of the polymer (90). Since a similar effect of tacticity has been observed during the nonoxidative thermal degradation of PVC (91,92), it seems that the mechanisms for the photochemical and nonphotochemical growth of polyenes may have similar steric requirements. Polyene formation in fluid media via a free-radical chain mechanism involving chlorine atoms has been shown to be highly unlikely on the basis of rela-tive reactivity considerations (91,92).

On the other hand, radicals are undoubtedly involved in the photodegradation of PVC under some experimental conditions. Recent ESR studies have provided evidence for the formation of alkyl and allyl-type radicals during the low-temperature UV irradiation of the polymer (43,72,87). Peroxy radicals were also observed when molecular oxygen was present (43,87). Other ESR work has shown convincingly that the radical $-CHCl-CH_2-\dot{C}H-CH_2-CHCl-$ results from the irradiation of PVC at liquid-nitrogen temperature (61,93) and is converted into a $-CH_2-\dot{C}Cl-CH_2-$ radical at -110°C (93).

One of the mechanisms proposed recently for the photooxidation of PVC involves the formation of methyl radicals from the ends of branches and chains (94). It is extremely difficult to understand how such radicals could be produced, since all of the saturated chain ends and branch ends identified thus far contain a terminal chloromethyl arrangement (see above).

The mechanisms for cross-linking and chain scission during the photodegradation of PVC have not yet been established. Possible cross-linking mechanisms include radical coupling and the addition of radicals to double bonds (43,87). Cross-linking is favored by irradiation at shorter wavelengths (56), and both cross-linking and chain scission have been found to occur under nitrogen as well as oxygen during irradiation of the polymer with (250-350)-nm light (30,31). Some possible chain-scission reactions are the Norrish type I and type II processes and the β cleavage of alkoxy radicals. Other possibilities are carbon-radical β scissions; for example, Reaction 13 ($\underline{n} \geq 0$).

$$-CH_2-(CH=CH)_{\underline{n}}-CHCl-CH_2-\dot{C}H- \longrightarrow$$

$$-CH_2-(CH=CH)_{\underline{n}}-\dot{C}HCl + CH_2=CH- \qquad (13)$$

Increases in $\underline{n}$ should facilitate such a process by increasing the resonance stabilization of the radical product.

Structures have not been determined for all of the oxygenated groups formed during the photooxidation of PVC. However, in one recent study, strong evidence was obtained for the formation of carboxyl groups at the scission points of polymer chains (64). Another investigation showed that the rate of appearance of color during PVC photooxidation could be correlated with the rates of formation of three types of carbonyl (69). The presence of conjugated carbonyl chromophores was suggested in order to account for this observation (69). Many workers have obtained infrared evidence for the formation of hydroxy and/or hydroperoxy groups during the photooxidation of the polymer (30,32, 43,44,49,50,51,57,58,62,64,69,85,86,87,94).

Singlet oxygen is more reactive than triplet oxygen toward polyene structures in PVC (95). Also, unlike triplet oxygen, singlet oxygen reacts preferentially with the shorter polyene sequences (95). The detailed chemistry of the $^1O_2$ oxidation is uncertain, and its relevance to the photooxidation of PVC remains to be established.

Photooxidation at 0°C in a sunshine weatherometer has been found to cause preferential destruction of the methylene groups in PVC (94). On the other hand, preferred removal of the chloromethylene groups was observed in an earlier photooxidation study carried out at 30°C with 253.7-nm irradiation (96). A possible explanation for this apparent contradiction is that the use of 253.7-nm light enhanced the relative importance of photoinitiation involving C-Cl homolysis.

Autoinhibition during the nonoxidative photodehydrochlorination (9,10,29,88,89,97) or photooxidation (25,49,52,59) of PVC films is usually attributed to the formation of a highly degraded surface layer that acts as a protective filter (9,10,25, 49,88,97). In addition, such a layer might retard the rate of photooxidation by inhibiting the diffusion of molecular oxygen (25). Polyenes (9,10,88,97) and carbonyl groups (58) do seem to be concentrated at sample surfaces under some conditions, at least (9,10,58,88,97). Nevertheless, autoinhibition has also been ascribed to the quenching of excited carbonyl groups by polyenyl radicals (52) and to the prevention of polyene-sensitized photoinitiation, owing to the formation of a "charge-transfer complex" (a carbenium chloride ion pair?) between the polyenes and HCl (59). In some situations, polyene formation is, indeed, inhibited by the presence of HCl (29), an effect which has been attributed to polyene destruction by HCl addition via a ground-state ionic mechanism (29). However, polyene bleaching by the action of HCl has been shown to occur photochemically also in a wavelength-dependent process (42).

In the early stages of reaction, HCl has been found to accelerate carbonyl formation during the photooxidation of PVC (59). This result has been attributed to an unprecedented HCl-catalyzed conversion of sec-peroxy radicals into hydroxy radicals and carbonyl groups (59).

In keeping with earlier observations (19,98), the non-oxidative thermal dehydrochlorination of PVC has been shown recently to be facilitated by preliminary photodegradation of the polymer (10,99). The thermal sensitivity enhancement increases with decreasing wavelength of irradiation (10) and undoubtedly results from the photolytic formation of thermally labile defect sites (10).

Spectrofluorimetric measurements have shown that degradation conditions can influence the nature of the fluorescent species in degraded PVC (100). On the basis of such measurements, a photooxidized polymer appeared to contain long polyenes, while a cyclohexadiene moiety seemed likely to be present in a sample that had been degraded thermally at 180°C in air (100). The latter result tends to support the involvement of cyclohexadiene

intermediates in the mechanism for the formation of benzene during the pyrolysis of PVC (91,92).
Photobleaching of the polyene sequences in degraded PVC occurs in either the presence or absence of oxygen (53). The process is wavelength-dependent, and its mechanism has been discussed (53). Benzophenone-sensitized photobleaching of PVC polyenes has also been studied recently (39,40). This reaction has the interesting property of being inhibited strongly by oxygen in THF but not in methylene chloride (39,40).
Finally, it is worth noting that, in response to triplet sensitization, simple acyclic allylic chlorides may be converted into the corresponding chlorocyclopropanes (101). Perhaps this reaction can also occur during the photodegradation of PVC.

## Concluding Remarks

Recent studies have provided much useful information about the photodegradation of PVC, but a thorough understanding of this subject has obviously not been achieved. Many of the pertinent data are contradictory for reasons not always apparent, although it is certain that the chemistry of the process depends very strongly on reaction variables. Clearly much work remains to be done in this very important field.

## Acknowledgments

The author is indebted to Drs. D. Braun, C. Decker, R. Gooden, and E. D. Owen for stimulating discussions regarding some of the unsolved problems that have been identified in this review.

## Abstract

The chemistry of the oxidative and nonoxidative photodegradation of poly(vinyl chloride) is reviewed with emphasis on work that has been published since the early 1970's. Topics covered include the nature of the photoinitiating species, the photoinitiation mechanism, and the structural consequences and reaction mechanism of the overall photodegradation process. Also included is a summary of recent studies on the determination of structural defects in poly(vinyl chloride) by carbon-13 NMR.

## Literature Cited

1. Close, L. G.; Gilbert, R. D.; Fornes, R. E. Polym.-Plast. Technol. Eng., 1977, 8, 177.
2. Owen, E. D. ACS Symp. Ser., 1976, 25, 208.
3. McKellar, J. F.; Allen, N. S. "Photochemistry of Man-Made Polymers"; Applied Science: London, 1979; pp. 95-102.
4. Wirth, H. O.; Andreas, H. Pure Appl. Chem., 1977, 49, 627.
5. Trozzolo, A. M. In "Polymer Stabilization"; Hawkins, W. L., Ed.; Wiley-Interscience: New York, 1972; Chapter 4.

6.  Braun, D.; Sonderhof, D. Third International Symposium on
    Poly(vinyl chloride), Preprints, Cleveland, Ohio, August
    1980, p. 14.
7.  Ratti, L.; Visani, F.; Ragazzini, M. Eur. Polym. J., 1973,
    9, 429.
8.  Kawai, W. Eur. Polym. J., 1974, 10, 805.
9.  Braun, D.; Wolf, M. Angew. Makromol. Chem., 1978, 70, 71.
10. Braun, D.; Wolf, M. Kunstst. Fortschrittsber., 1976, 2, 13.
11. Kawai, W.; Ichihashi, T. J. Polym. Sci., Polym. Chem. Ed.,
    1974, 12, 201.
12. Owen, E. D.; Bailey, R. J. J. Polym. Sci., Part A-1, 1972,
    10, 113.
13. Golub, M. A. J. Phys. Chem., 1971, 75, 1168.
14. Harriman, A.; Rockett, B. W.; Poyner, W. R. J. Chem. Soc.,
    Perkin Trans. 2, 1974, 485.
15. Hirayama, S.; Foster, R. J.; Mellor, J. M.; Whitling, P. H.;
    Grant, K. R.; Phillips, D. Eur. Polym. J., 1978, 14, 679.
16. Wagner, P. J.; Sedon, J. H.; Lindstrom, M. J. J. Am. Chem.
    Soc., 1978, 100, 2579.
17. Wagner, P. J.; Sedon, J. H. Tetrahedron Lett., 1978, 1927.
18. Wagner, P. J.; Lindstrom, M. J. "Abstracts of Papers",
    177th National Meeting of the American Chemical Society,
    Honolulu, Hawaii, April 1979; American Chemical Society:
    Washington, D.C., 1979; ORGN 210.
19. Kenyon, A. S. Natl. Bur. Stand. (U.S.) Circ., 1953, 525, 81.
20. Golub, M. A. J. Am. Chem. Soc., 1969, 91, 4925.
21. Golub, M. A. J. Am. Chem. Soc., 1970, 92, 2615.
22. Kagiya, V. T.; Takemoto, K.; Hagiwara, M. J. Appl. Polym.
    Sci.: Appl. Polym. Symp., 1979, 35, 95.
23. Owen, E. D.; Williams, J. I. J. Polym. Sci., Polym. Chem.
    Ed., 1973, 11, 905.
24. Harriman, A.; Rockett, B. W. J. Chem. Soc., Perkin Trans. 2,
    1974, 1235.
25. Braun, D.; Kull, S. Angew. Makromol. Chem., 1980, 86, 171.
26. Zweig, A.; Henderson, Jr., W. A. J. Polym. Sci., Polym.
    Chem. Ed., 1975, 13, 717.
27. Heskins, M.; Reid, W. J.; Pinchin, D. J.; Guillet, J. E.
    ACS Symp. Ser., 1976, 25, 272.
28. Kawai, W.; Ichihashi, T. J. Polym. Sci., Polym. Chem. Ed.,
    1974, 12, 1041.
29. Gibb, W. H.; MacCallum, J. R. Eur. Polym. J., 1974, 10, 533.
30. Balandier, M.; Decker, C. Eur. Polym. J., 1978, 14, 995.
31. Decker, C.; Balandier, M. IUPAC 26th International Symposium
    on Macromolecules, Preprints, Mainz, Federal Republic of
    Germany, September 1979, Vol. I, p. 588.
32. Decker, C. Second International Conference on Advances in
    the Stabilization and Controlled Degradation of Polymers,
    Preprints, Luzern, Switzerland, June 1980.
33. Turro, N.J. "Modern Molecular Photochemistry"; Benjamin/
    Cummings: Menlo Park, California, 1978; p. 292.

34. Reference 33, p. 181.
35. Reinisch, R. F.; Gloria, H. R.; Androes, G. M. In "Photochemistry of Macromolecules"; Reinisch, R. F., Ed.; Plenum: New York, 1970; p. 185.
36. Phillips, R. W.; Volman, D. H. J. Am. Chem. Soc., 1969, 91, 3418.
37. Gibb, W. H.; MacCallum, J. R. Eur. Polym. J., 1974, 10, 529.
38. Kohn, P.; Marechal, C.; Verdu, J. Anal. Chem., 1979, 51, 1000.
39. Owen, E. D.; Pasha, I. IUPAC 26th International Symposium on Macromolecules, Preprints, Mainz, Federal Republic of Germany, September 1979, Vol. I, p. 592.
40. Owen, E. D.; Pasha, I. Am. Chem. Soc., Div. Org. Coat. Plast. Chem., Pap., 1980, 42, 724.
41. Gibb, W. H.; MacCallum, J. R. Eur. Polym. J., 1973, 9, 771.
42. Owen, E. D.; Williams, J. I. J. Polym. Sci., Polym. Chem. Ed., 1974, 12, 1933.
43. Rabek, J. F.; Canbäck, G.; Lucky, J.; Ranby, B. J. Polym. Sci., Polym. Chem. Ed., 1976, 14, 1447.
44. Rabek, J. F.; Canbäck, G.; Ranby, B. J. Appl. Polym. Sci.: Appl. Polym. Symp., 1979, 35, 299.
45. Mori, F.; Koyama, M.; Oki, Y. Angew. Makromol. Chem., 1979, 75, 223.
46. Starnes, Jr., W. H.; Schilling, F. C.; Abbas, K. B.; Cais, R. E.; Bovey, F. A. Macromolecules, 1979, 12, 556.
47. Starnes, Jr., W. H.; Schilling, F. C.; Abbas, K. B.; Cais, R. E.; Bovey, F. A. Polym. Prepr., Am. Chem. Soc., Div. Polym. Chem., 1979, 20(1), 653.
48. Verdu, J.; Michel, A.; Sonderhof, D. Eur. Polym. J., 1980, 16, 689.
49. Scott, G.; Tahan, M.; Vyvoda, J. Eur. Polym. J., 1978, 14, 1021.
50. Scott, G. Adv. Chem. Ser., 1978, 169, 30.
51. Scott, G. Polym.-Plast. Technol. Eng., 1978, 11, 1.
52. Marechal, J. C. J. Macromol. Sci., Chem., 1978, 12, 609.
53. Owen, E. D.; Read, R. L. J. Polym. Sci., Polym. Chem. Ed., 1979, 17, 2719.
54. Gibb, W. H.; MacCallum, J. R. J. Polym. Sci., Polym. Symp., 1973, 40, 9.
55. Ng, H. C.; Guillet, J. E. Macromolecules, 1978, 11, 937.
56. Kamal, M. R.; El-Kaissy, M. M.; Avedesian, M. M. J. Appl. Polym. Sci., 1972, 16, 83.
57. Rabek, J. F.; Shur, Y. J.; Ranby, B. J. Polym. Sci., Polym. Chem. Ed., 1975, 13, 1285.
58. Mori, F.; Koyama, M.; Oki, Y. Angew. Makromol. Chem., 1977, 64, 89.
59. Verdu, J. J. Macromol. Sci., Chem., 1978, 12, 551.
60. Foster, R. J.; Whitling, P. H.; Mellor, J. M.; Phillips, D. J. Appl. Polym. Sci., 1978, 22, 1129.
61. Rabek, J. F.; Skrowronski, T. A.; Ranby, B. Polymer, 1980, 21, 226.

62.  Rabek, J. F.; Canbäck, G.; Rånby, B. J. Appl. Polym. Sci.,
     1977, 21, 2211.
63.  Mori, F.; Koyama, M.; Oki, Y. Angew. Makromol. Chem., 1979,
     75, 123.
64.  Scott, G.; Tahan, M. Eur. Polym. J., 1975, 11, 535.
65.  Scott, G.; Tahan, M.; Vyvoda, J. Eur. Polym. J., 1979, 15,
     51.
66.  Cooray, B. B.; Scott, G. Polym. Degradation Stab., 1980, 2,
     35.
67.  Braun, D.; Kull, S. Angew. Makromol. Chem., 1980, 87, 165.
68.  Bellenger, V.; Verdu, J.; Carette, L. B. Third International
     Symposium on Poly(vinyl chloride), Preprints, Cleveland,
     Ohio, August 1980, p. 297.
69.  Scott, G.; Tahan, M. Eur. Polym. J., 1977, 13, 989.
70.  Rabek, J. F. ACS Symp. Ser., 1976, 25, 255.
71.  Rånby, B.; Rabek, J. F. J. Appl. Polym. Sci.: Appl. Polym.
     Symp., 1979, 35, 243.
72.  Joffe, Z.; Rånby, B. J. Appl. Polym. Sci.: Appl. Polym.
     Symp., 1979, 35, 307.
73.  Scott, G.; Tahan, M. Eur. Polym. J., 1977, 13, 997.
74.  Bovey, F. A.; Abbas, K. B.; Schilling, F. C.; Starnes, Jr.,
     W. H. Macromolecules, 1975, 8, 437.
75.  Starnes, Jr., W. H.; Schilling, F. C.; Abbas, K. B.; Plitz,
     I. M.; Hartless, R. L.; Bovey, F. A. Macromolecules, 1979,
     12, 13.
76.  Starnes, Jr., W. H.; Schilling, F. C.; Plitz, I. M.;
     Hartless, R. L.; Bovey, F. A. Polym. Prepr., Am. Chem. Soc.,
     Div. Polym. Chem., 1978, 19(2), 579.
77.  Starnes, Jr., W. H.; Hartless, R. L.; Schilling, F. C.;
     Bovey, F. A. Adv. Chem. Ser., 1978, 169, 324.
78.  Starnes, Jr., W. H.; Hartless, R. L.; Schilling, F. C.;
     Bovey, F. A. Polym. Prepr., Am. Chem. Soc., Div. Polym.
     Chem., 1977, 18(1), 499.
79.  Abbas, K. B.; Bovey, F. A.; Schilling, F. C. Makromol.
     Chem., Suppl., 1975, 1, 227.
80.  Starnes, Jr., W. H.; Schilling, F. C.; Plitz, I. M.; Cais,
     R. E.; Freed, D. J.; Bovey, F. A. Third International
     Symposium on Poly(vinyl chloride), Preprints, Cleveland,
     Ohio, August 1980, p. 58.
81.  Starnes, Jr., W. H.; Schilling, F. C.; Plitz, I. M.; Cais,
     R. E.; Freed, D. J.; Bovey, F. A., manuscript in prepara-
     tion.
82.  Bovey, F. A.; Schilling, F. C.; Starnes, Jr., W. H. Polym.
     Prepr., Am. Chem. Soc., Div. Polym. Chem., 1979, 20(2),
     160.
83.  Starnes, Jr., W. H. Adv. Chem. Ser., 1978, 169, 309; re-
     ferences cited therein.
84.  Starnes, Jr., W. H. Polym. Prepr., Am. Chem. Soc., Div.
     Polym. Chem., 1977, 18(1), 493; references cited therein.

85.   Mori, F.; Koyama, M.; Oki, Y. Angew. Makromol. Chem., 1979,
      75, 113.
86.   Kwei, K.-P. S. J. Polym. Sci., Part A-1, 1969, 7, 1075.
87.   Ranby, B.; Rabek, J. F.; Canback, G. J. Macromol. Sci.,
      Chem., 1978, 12, 587.
88.   Braun, D.; Kull, S. Angew. Makromol. Chem., 1980, 85, 79.
89.   Gibb, W. H.; MacCallum, J. R. Eur. Polym. J., 1972, 8,
      1223.
90.   Mitani, K.; Ogata, T. J. Appl. Polym. Sci., 1974, 18, 3205.
91.   Starnes, Jr., W. H.; Edelson, D. Macromolecules, 1979, 12,
      797; references cited therein.
92.   Starnes, Jr., W. H.; Edelson, D. Am. Chem. Soc., Div. Org.
      Coat. Plast. Chem., Pap., 1979, 41, 505; references cited
      therein.
93.   Yang, N.-L.; Liutkus, J.; Haubenstock, H. Polym. Prepr., Am.
      Chem. Soc., Div. Polym. Chem., 1979, 20(2), 195.
94.   Mori, F.; Koyama, M.; Oki, Y. Angew. Makromol. Chem., 1978,
      68, 137.
95.   Rabek, J. F.; Ranby, B.; Ostensson, B.; Flodin, P. J. Appl.
      Polym. Sci., 1979, 24, 2407.
96.   Kwei, K.-P. S. J. Polym. Sci., Part A-1, 1969, 7, 237.
97.   Gibb, W. H.; MacCallum, J. R. Eur. Polym. J., 1971, 7, 1231.
98.   Druesedow, D.; Gibbs, C. F. Natl. Bur. Stand. (U.S.) Circ.,
      1953, 525, 69.
99.   Gupta, V. P.; St. Pierre, L. E. J. Polym. Sci., Polym. Chem.
      Ed., 1979, 17, 931.
100.  Owen, E. D.; Read, R. L. Eur. Polym. J., 1979, 15, 41.
101.  Cristol, S. J.; Daughenbaugh, R. J. J. Org. Chem., 1979,
      44, 3434; references cited therein.

RECEIVED November 13, 1980.

# The Roles of Hydrogen Chloride in the Thermal and Photochemical Degradation of Polyvinyl Chloride

ERYL D. OWEN

Chemistry Department, University College, Cardiff, Wales, U.K.

Poly(vinylchloride), (PVC), has been a polymer of consider-able commercial importance for about forty years. The wide variety of applications of the polymer has meant that a considerable amount of research has been conducted concerning all aspects of its preparation, processing and properties. Foremost amongst the problems which still remain to be solved however are the reasons for the thermal instability which present problems to the technologist during the processing which takes place within the temperature range 200-250°C. Closely related problems concern the photochemical instability at much lower temperatures which severely limits the potentially large outdoor applications of the polymer as well as the deterioration which occurs on exposure to high energy radiation. Rates of degradation in most cases can be reduced to a level which is acceptable commercially by the incorporation of stabilizers into the polymer but those currently in use are by no means ideal and many aspects of the problem remain.

The primary product which arises from the degradation of PVC, whether induced thermally, photochemically or by high energy radiation, is a distribution of conjugated polyene sequences of various lengths produced by a dehydrochlorination process which may be written:-

$$\sim(CH_2 \cdot CHCl)_n \longrightarrow \sim(CH=CH)_n\sim + nHCl \text{----}(1)$$

The details of the process by which the reaction is initiated and propagated in each case have been discussed many times but are still far from being completely understood. It is generally agreed that the presence of various structural features in the polymer, which occur to different extents depending on the polymerization conditions, are important particularly since recent improvements in analytical techniques have made it easier to identify them and quantify their low concentration levels with greater certainty. Although deterioration of any of the physical or mechanical properties of PVC is undesirable much of

0097-6156/81/0151-0217$05.50/0

the effort has been directed towards prevention of the
coloration which appears as the lengths of the polyene sequences
increase to the point where their absorptions extend into the
visible region of the spectrum. For this reason measurements
of absorption in the ultra-violet and visible regions have been
widely used as an indication of the extents of degradation. It
has become clear from our earlier work however that polyenes
formed in theprimary step are extremely reactive, forming
secondary products which may adversely affect the stability of
the polymer by sensitizing further degradation. The relative
importance of several separate aspects of the degradation
including initiation, propagation, termination and length of
polyene sequences formed seem also to be influenced by the
presence of one of the primary reaction products namely hydrogen
chloride, (HCl), and it is only relatively recently that the
extent of the different and divergent roles played by HCl has
become apparent.

1.    Summary of Early Work
     Almost as soon as PVC began to be produced as a commercial
polymer, technologists suspected that the thermal degradation
which occurs during the processing is catalysed by the HCl
liberated in an autocatalytic process. One of the main reasons
for this belief was the obvious effect of acid acceptors in
stabilizing the polymer towards thermal degradation. The idea
was challenged initially following the failure of some workers,
notably Arlmann (1) and Druesdow and Gibb (2) to detect any
catalytic effect of HCl. With the benefit of hindsight it is
not surprising that these experiments produced negative results,
since, by present standards, they were relatively insensitive.
In one case (1) a carrier gas technique was involved which used
nitrogen, oxygen or air, and which was therefore complicated
by the fact that oxygen has a positive effect on the
dehydrochlorination rate.   Reinterpretation of these results
in the light of this information indicates that a small
accelerating effect was apparent. The more sensitive technique
used by Druesedow  and Gibb (2), which involved comparing the
rate of dehydrochlorination while the carrier gas flowed with
that when the flow was interrupted and the HCl allowed to
accumulate, also gave negative results. The result was widely
accepted due partly to the difficulty of making accurate
measurements of HCl evolved but equally because of the
difficulty of fitting HCl catalysis into the current ideas of
the mechanism by which PVC underwent thermal degradation.
     Since that time increasing numbers of experiments have
been carried out using analytical techniques of improved
sensitivity and most workers now agree that there is overwhelming
evidence that HCl does catalyze the thermal decomposition of
PVC both in the presence of oxygen and under inert conditions.
Among the first group of workers whose results showed a positive

accelerating effect of HCl were Riecke, Grimm and Mücke (3) and
a most convincing experiment was described by Talamini, Cinque
and Palma (4) who degraded solid PVC in a vacuum apparatus
in which HCl was removed, condensed, isolated and measured in a
gas burette. In a subsequent paper the same group of workers
(5) described the perhaps even more significant result that
when accumulated HCl was removed, the increasing rate of
dehydrochlorination was reduced to a constant value but one
which was higher than that observed for systems in which the
HCl was never allowed to accumulate.

General acceptance that HCl does have a catalytic effect
was not sufficient impetus to resolve the problem of the
degradation mechanism which was assumed by most workers to be
a radical type, similar to that suggested by Winkler (6) in
spite of some features which suggested strongly that at least
some degree of charge separation must be involved. In
addition, Braun and Bender (7) showed that when PVC was
thermally degraded at 160-200°C in a series of solvents which
included ethyl benzoate, the rate of dehydrochlorination followed
a first order rate law and the rate was increased by the
presence of HCl or of oxygen. In contrast, the presence of the
free radical initiators azoisobutyronitrile, (AIBN), or
tetramethoxybenzopinacol, (TMBP), had no such effect and on the
basis of this evidence a unimolecular elimination, (eqn.2),
was suggested.

$$\sim CH_2 - \overset{\overset{\displaystyle Cl}{|}}{CH} - \overset{\overset{\displaystyle H}{|}}{CH} - CHCl \sim \quad \xrightarrow{\text{-HCl}} \quad \sim CH_2 - CH = CH - CHCl \sim$$

$$\text{-------(2)}$$

A further complication was reported by Van der Ven and deWit (8)
who, using a sensitive conductometric technique for measuring
the HCl evolved, showed that the accelerating effect was
greater for films than for powders. A similar effect was
reported by Thallmeier and Braun (9).

These remarks represent only the barest outline of at least
two aspects of PVC degradation which have been the focus of
attention for several years and remain incompletely understood
namely the mechanism involved and the related problem of the
involvement of HCl. Several excellent reviews give more
comprehensive summaries of the earlier work (10, 11, 12). More
recent work has made it clear that under appropriate conditions
the presence of HCl can affect the initiation, propagation and
termination steps as well as influencing the distribution of
polyene sequence lengths. In addition it can undergo photochemical addition reactions with the polyenes, i.e. the reverse of the
dehydrochlorination process, as well as forming colored
polyene/HCl complexes. These various possibilities will be
considered in turn.

2.   Effect of HCl on Initiation, Propagation and Termination
     Steps
     A detailed quantitative description of the effect of HCl on
the three distinct phases of the degradation process requires as
an obvious prerequisite some knowledge of the nature of the
mechanisms in the absence of HCl.  An aspect of the problem on
which a great deal of attention has been concentrated recently is
the relative concentration of the various defect structures in the
polymer and their different roles in influencing the type of
mechanism which will occur.  Although a majority consensus now
seems to be in favor  of a molecular elimination with some ionic
contribution, this is by no means a unanimous view.  Recent
results by some workers (13, 14, 15) appear to provide evidence
which favors  a radical mechanism while ionic mechanisms
seem to explain best results of solvent studies both of PVC and
model allylic compounds.  It seems certain therefore that no
single mechanism can account for all the data obtained under
various conditions of degradation which are encountered and
indeed dual mechanisms may be operative even in a single system
(16).

a)   Initiation
     While emphasising the difficulty of designing experiments
which can discriminate between the effects of HCl on the initiation, propagation or termination steps,  Hjertberg and Sorvik (17)
concluded that initiation of the thermal degradation of PVC at
190°C was catalysed by HCl for atmospheres which contained
10%-40% HCl.  Other workers on the other hand (18, 19, 20) have
concluded that the rate of initiation is unaffected by HCl or that
initiation arising from chloroallylic groups present is the only
catalyzed step.  If the non allylic initiation is catalyzed by
HCl then the most likely mechanism would involve a cyclic transition state of the type proposed by Imomoto (31), (equation 3).

$$\sim CH_2 - CH - CH - CHCl \sim \longrightarrow \sim CH_2 - CH{=}CH - CHCl \sim + \ 2HCl$$

----(3)

Most of the other possibilities which have been suggested require
the presence of some unsaturation the most often quoted being the
allylic C-Cl bond mentioned already.  In all those cases at
least some degree of charge separation is implied e.g. by
Van der Ven (8), (equation 4), Morikawa, (21), (equation 5) or
Rasuvaev, (22), (equation 6).

$$Cl^- + \sim CH{=}CH - CH - CH_2 \sim \longrightarrow \sim (CH{=}CH)_2 \sim + \ HCl + Cl^-$$

(or $HCl_2^-$)                    Cl                              (or $HCl_2^-$)

----(4)

$$\sim (CH=CH)_n - CH - CH \sim \longrightarrow \sim (CH=CH)_{n+1} \sim + \ 2HCl \ ----(5)$$

$$\delta+H - Cl_\delta --- H \quad Cl$$

$$\sim (CH=CH)_n - CH - CH \sim \longrightarrow \sim (CH=CH)_{n+1} \sim + \ 2HCl \ ----(6)$$

$$Cl \quad H$$

$$H - Cl$$

b)   Propagation

The effect of HCl on the propagation step appears  less controversial since most workers are agreed that its presence does increase the rate of propagation.  The mechanism has some degree of ionic character and may be represented in an extreme form by equation (7).

$$\sim (CH=CH)_n - CHCl - CH_2 \sim \longrightarrow \sim (CH=CH)_n - \overset{+}{CH} - CH_2 \sim$$

$$+ \ HCl \qquad\qquad\qquad\qquad + \ HCl_2^-$$

$$\downarrow$$

$$\sim (CH=CH)_{n+1} \sim + \ 2HCl$$

$$----(7)$$

It may be regarded as an extension of the allyl activated initiation step, becoming more facile as the value of n increases. At this point a semantic question of what should be regarded as initiation and where propagation begins becomes apparent but is most easily resolved by considering the activation energies or rate constants of the reactions involved.  Hjertberg and Sorvik (17) have shown that the propagation rate increases to a maximum value somewhere in the range 0-10% HCl in the atmosphere and most workers are agreed that the rate of catalysed propagation exceeds that of initiation by some orders of magnitude.  The rate of propagation and consequent polyene sequence length is also facilitated by syndiotactic arrangements since polymers with high syndiotactic content produce abnormally long polyene sequences (23).  It has also been suggested (22) that for such long polyenes, triplet excited states may be intermediates since they may be populated even at room temperature (24).

c)   Termination

Most workers are in agreement with the suggestion of Marks (25) and others (22) that the increased formation of gel content with increasing HCl concentration is due to an increased rate of termination by inter-chain reactions leading to cross links.   Hjertberg and Sorvik though (17) believe that the termination rate constant is unaffected but that the increased extent of molecular enlargement only reflects the increased extent of dehydrochlorination. Although many suggestions can be made regarding the nature of the termination process and the related reasons for the relatively short polyene sequence lengths, this aspect of the problem is not well understood. One additional complicating factor namely photochemically induced cross-linking will be discussed in section 3.

In a very recent paper (47), Amer and Shapiro concluded that the thermal degradation of PVC powder in the temperature range 170-210°C proceeded in such a way that HCl catalysis was an integral part of the overall process. They proposed a unified mechanism consisting of three steps namely random generation of a single double bond in the cis configuration, 1,4 elimination of HCl via a six centred transition state yielding a polyene, then HCl catalyzed isomerization of the polyene formed to regenerate the initial structure.

The effect of HCl on the photolysis of PVC films has been shown (30) to be complicated by the formation of a thin surface layer of highly absorbing polyenes. Such a layer may act as an effective filter and reduce dramatically the amount of light which can penetrate to the interior of the film and hence protect the bulk from further degradation. On the other hand Decker has shown (48) that when PVC films were irradiated, (250-350 nm) either in an atmosphere of pure nitrogen or pure oxygen, HCl was evolved at an increasing rate. He concluded that the auto accelerating process arises exclusively from the increased absorption of light by the polyenes formed which photo-sensitize further degradation and that the quantum yield of HCl formation was constant ($\emptyset$ = 0.11 in nitrogen and 0.15 in the presence of oxygen). Verdu (49) on the other hand, in an earlier spectrophotometric study of the rates of photochemical formation of carbonyl groups and polyene sequences in films of different thicknesses, had concluded that the initial autocatalytic effect of HCl was due to a catalyzed unimolecular decomposition of peroxy radicals forming carbonyl groups and hydroxyl radicals. The autoinhibiting effect which was apparent in the later stages, particularly for thicker films, was attributed to the formation of charge transfer complexes between HCl and polyenes. These complexes were assumed to be photochemically inert as far as further dehydrochlorination of the polymer was concerned but may be involved in the reverse process of re-addition of HCl to the polyenes. The balance between the two effects depends on the ease of diffusion of HCl out of the film.

3.  Addition of HCl to Polyenes

The thermally activated addition of HCl to simple olefins in polar solvents has been well known for many years. Corresponding additions in the gas phase have high activation energies and over temperature ranges at which the reaction can conveniently be measured, the equilibrium lies largely on the side of elimination rather than addition.  On the other hand additions proceed smoothly in aprotic polar solvents between -80 and +80ºC to produce products which have Markovnikov type orientation.  The work of Pocker in this field (25) has shown for example that HCl adds to 2 methyl-1-butene or 2 methyl-2-butene in nitromethane solvent and that unionized HCl is the dominant proton donor.  The transfer of proton to the olefin, which is the rate controlling step, is assisted by a second molecule of acid which hydrogen bonds to the developing chloride ion forming $HCl_2^-$.  This and other work indicates that HCl is largely undissociated in nitromethane for $[HCl] > \sim 0.015$ M and that there is little association either.  There is evidence that a corresponding addition occurs to olefins in thermally degraded PVC.  Results carried out in a variety of solvents (26) are consistent with elimination of HCl occurring by a $\beta$ -elimination of the $E_1$ type favored  by polar solvents.  The same authors showed that at least in nitrobenzene containing dissolved HCl, the reverse reaction, i.e. addition of HCl, takes place.  The fact that this may be interpreted as a retardation of the degradation process may have contributed to the confusion which has arisen and emphasizes the care which must be taken to  disentangle the possible catalytic effect of HCl when concurrent addition of HCl to the polyenes is possible.

The photochemical addition of HCl to polyenes in degraded PVC has been shown (27) to proceed smoothly in the solid state.

Films cast from thermally degraded PVC were irradiated with light from a G.E.C. 250-W medium-pressure mercury arc filtered through a Pyrex glass disk (10% transmission at 308 nm and 2% at 299 nm) in the presence of various pressures of HCl gas.  On photolysis, the absorbance between 270 nm and about 415 nm decreased but at wavelengths shorter than 270 nm and longer than 415 nm the absorbance increased.  Figure 1 is a typical series of spectra after different times of photolysis when the pressure of HCl was 50 torr.  The initial rate of the bleaching reaction (270-415 nm), depended on the pressure of HCl (Fig. 2).  Figure 3 shows that the overall extent of the change appeared to reach a limiting value at high pressures of HCl.  In the absence of HCl a very much slower bleaching reaction was observed.

When similar experiments were carried out with the use of narrow bands of irradiation isolated by metal interference filters, the results were essentially the same, except that the maximum decrease in absorbance occurred at a wavelength which depended on that of the exciting light (Fig. 4).  At wavelengths longer than 368 nm no bleaching was observed.  The higher the pressure of HCl at a particular wavelength, the broader the extent of the photobleaching, i.e.  the further it extended into the visible and ultraviolet regions.

Experiments in which a silica disk was the only filter used, so that the film was photolyzed with the whole visible and ultraviolet output of the lamp, showed that under these conditions the bleaching reaction was accompanied by a background photo-

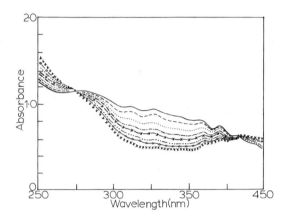

Journal of Polymer Science, Polymer Chemistry Edition

*Figure 1.    Absorption spectra of thermally degraded PVC film irradiated (λ > 300 nm) in an atmosphere containing 50 torr HCl for various lengths of time: (———) 0; (– – –) 5 min; (· · ·) 20 min; (· – ·) 40 min; (– ✕ –) 60 min; (· · · –) 100 min; (○–○) 160 min; (✕) 220 min (27)*

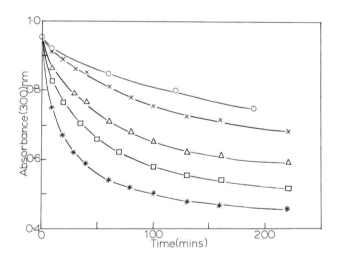

Journal of Polymer Science, Polymer Chemistry Edition

*Figure 2.    Absorbance at 300 nm of thermally degraded PVC films on irradiation (λ > 300 nm) in the presence of various pressures of HCl: (○) 0; (✕) 50 torr; (△) 150 torr; (□) 300 torr; (*) 760 torr (27)*

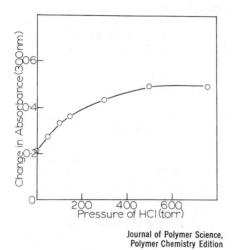

Journal of Polymer Science,
Polymer Chemistry Edition

*Figure 3. Dependence on the HCl pressure of the extent of the photobleaching of thermally degraded PVC films, measured at 300 nm (27)*

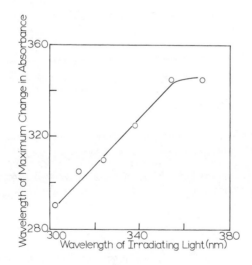

Journal of Polymer Science, Polymer Chemistry Edition

*Figure 4. Relation between the wavelength where maximum photobleaching occurs and the wavelength of the irradiating light (27)*

degradation reaction which was characterized by an increase in absorbance at all wavelengths. This reaction was slow compared with the bleaching reaction and began to become apparent only when the absorbance following the bleaching reaction had reached a minimum value after about 70 min (Fig. 5). Thereafter the photodegradation continued at a rate which was independent of the HCl pressure.

Absorptions in the region 270-415 nm would be expected to contain contributions from polyene sequences of the general structure $\sim(-CH=CH-)_n\sim$ , where n = 3-15; since the distribution of these species may be expected to vary according to the details of the method by which the degradation was produced, a considerable complexity of behavior is to be expected. The decrease in the absorbance at a particular wavelength with a corresponding increase at shorter wavelengths ( < 270) can most obviously be attributed to the reaction of a polyene sequence $\sim(-CH=CH-)_n\sim$ with HCl with the formation of a shorter sequence having an absorption at shorter wavelength. The fact that the rate of reaction depended on the HCl pressure and appeared to reach a limiting value (50% of its initial intensity for 760 torr HCl) suggested that the process may be reversible and may occur in two ways,

$$\sim(CH=CH)_n\sim + HCl \underset{h\nu'}{\overset{h\nu}{\rightleftharpoons}} \sim(CH=CH)_{n-1}\overline{\phantom{-}} CH_2 - CHCl\sim \quad ----(8)$$

or

$$\sim(CH=CH)_n\sim + HCl \underset{h\nu'}{\overset{h\nu}{\rightleftharpoons}}$$

$$\sim(CH=CH)_x - CH_2 - CHCl - (CH=CH)_{n-x-1}\sim \qquad ----(9)$$

depending on whether the addition of HCl takes place near the middle or at the end of a polyene sequence. The possibility that reactions (8) and (9) may be reversible is supported by the fact that the photosensitized elimination of HCl from alkyl halides has been shown (28) to occur, although it is not clear whether excited electronic states or vibrationally excited ground states are involved.

Using 1,8-diphenyloctatetra-1,3,5,7-ene, (DOT), as a model compound either in dilute, ($\sim 10^{-5}M$), hexane or ethanol solutions or incorporated into a film of undegraded PVC confirmed that in the presence of HCl it underwent a photochemical reaction which resembled that of the polyenes in thermally degraded PVC. The results indicated that the initial rates of reactions proceeding in either solvent showed a second order dependence on HCl pressure and that the reaction was considerably slower in ethanol than in hexane. Further, when cast in PVC films, the characteristic absorption maxima of DOT were shifted about 16nm to longer wavelengths compared with their absorption in hexane and there was a linear dependence of initial photobleaching rate on HCl pressure.

When a particular range of polyene sequences in PVC is selected by irradiating with a narrow band of frequencies, it is not surprising that the maximum amount of photobleaching is observed at a wavelength which is close to that of the irradiating light.

The fact that irradiation of the degraded films without using the Pyrex filter caused a superimposed degradation reaction to take place at a rate which was in excess of that for films which did not contain some degraded polymer suggests that another sensitized process should be included, namely, energy transfer from degraded to undegraded PVC:

$$\sim (CH=CH)_n^* + \sim (CH_2 - CHCl)_m \sim$$
$$\longrightarrow \sim (CH=CH)_n \sim + \sim (CH_2 - CHCl)_m^* \sim$$
$$\downarrow$$
$$\sim (CH_2 - CHCl)_{m-1}(CH=CH) \sim + HCl \quad --(10)$$

as well as intrapolyene transfer

$$\sim (CH=CH)_m^* + \sim (CH=CH)_n \sim \longrightarrow \sim (CH=CH)_m \sim$$
$$+ \sim (CH=CH)_n^* \sim \qquad ---(11)$$

The same conclusions were reached but from the opposite direction by Gibb and MacCallum (29) following experiments in which they irradiated non degraded PVC films (a) with continuing nitrogen purge, (b) starting with pure nitrogen and retaining the evolved HCl in the cell and (c) filling the cell with one atmosphere of HCl before irradiating. The results showed that the increases in absorbance due to formation of polyenes was reduced considerably by the presence of HCl. They suggested a reaction scheme which was the same as that described by equations 10 and 11 and favored an ionic mechanism. Kinetic treatments in this case are complicated by the fact that photodegradation is confined to a small surface layer only a few microns thick and although the film may appear highly colored the degraded material represents only a small proportion of the total leaving the bulk unaffected (30). Extent of reaction therefore depends on the balance between the rate constants for the individual processes and the rate of diffusion of HCl the importance of which has been mentioned already.

4. Effect of HCl on Polyene Sequence Lengths Produced by Thermal Degradation

Many workers (17, 33) have noticed that the presence of HCl

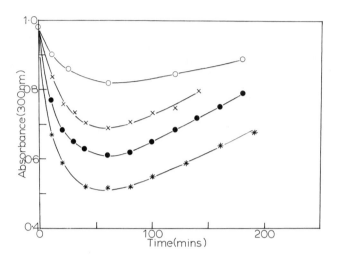

Journal of Polymer Science, Polymer Chemistry Edition

*Figure 5. Absorbance at 300 nm of thermally degraded PVC films irradiated with unfiltered light from a high-pressure mercury–xenon lamp in the presence of various pressures of HCl (27). (O–O) 0; (×–×) 50; (●–●) 300; (\* – \*) 760 torr.*

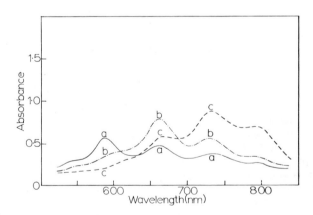

Journal of Applied Polymer Science

*Figure 6. Absorption spectra of dichloromethane solutions of chemically degraded PVC containing trifluoroacetic acid (0.1M) (a) after mixing; (b) after 20 min; and (c) after 200 min (38)*

affects the distribution of polyene sequence lengths and though they have not been unanimous in deciding in which direction, it appears in the main that the presence of HCl shifts the distribution towards longer sequences. In the absence of HCl the distribution is usually similar to that found by Bengough (34) but under some conditions (35, 36) the distribution may be shifted drastically in favor of longer sequences resulting in a pronounced absorption maximum around 450-500 nm. Palma has shown (37) that both kinds of distribution can be obtained depending on whether the HCl produced is allowed to remain in the neighborhood of the degrading polymer or rigorously removed. Similar differences have been noticed between polymer films and polymer powder and even between thin and thick layers of powder.

In order to investigate how HCl affects the polyene sequences we (38) have investigated the effect of trifluoroacetic acid (TFA) on the polyenes introduced chemically into PVC using the method of Shindo (39). TFA was chosen since its concentration in solution can be controlled and set at higher levels than is possible with HCl thus allowing the possibility of producing high concentration of intermediates which can be detected.

When dichloromethane (DCM) solutions of the polyenes which had been prepared as described (39) were added to DCM solutions of TFA new species were formed which had strong absorptions in the region 500-850 nm. Figure 6 shows some typical spectra for such a solution, (a) immediately after mixing, (b) after a further 20 minutes and (c) after 200 minutes. In spectrum (a) clearly defined maxima are visible at 590, 660, 730 and 790 nm the intensities of which change with time in a way which indicates that they are inter-related (figure 7). As $A_{590}$ decreases, $A_{660}$ increases until it in turn begins to decrease and is replaced by $A_{730}$ which finally gives way to $A_{790}$. The obviously large overlap of the adjacent maxima makes detailed kinetic analysis difficult but some qualitative conclusions can be drawn. Figure 8 shows the effect of varying the TFA concentration on the rate of change of the absorbance at 790 nm. Similar changes are observed at the other wavelengths which correspond to the absorption maxima. This data is summarized in figures 9 and 10 which show the maximum absorbance reached and the maximum rate of increase of absorbance for each wavelength at which an absorbance maximum occurs for different TFA concentration. When these experiments using TFA were repeated using chloroform or ortho-dichlorobenzene solvent instead of DCM similar results were obtained with only small differences in the positions and relative intensities of the absorption maxima but no such absorptions were detected when the solvents were tetrahydrofuran (THF) or cyclohexanone (CH).

In all the solvents studied the species responsible for the color with TFA were extremely photosensitive and were bleached in a few seconds when irradiated with unfiltered light from a medium pressure mercury lamp.

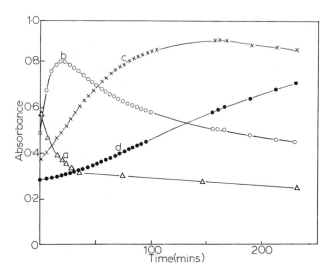

*Figure 7.    Changes in the absorbance at (a) 590, (b) 660, (c) 730, and (d) 790 nm for dichloromethane solutions of chemically degraded PVC containing trifluoro-acetic acid (0.3M) (38)*

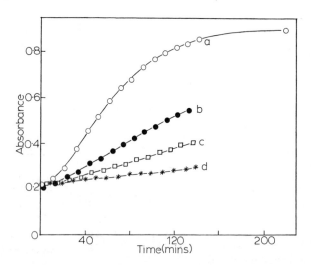

*Figure 8.    Effect of changing trifluoroacetic acid concentration on the rate of increase in absorbance at 790 nm for dichloromethane solutions of chemically degraded PVC: (a) 0.1M; (b) 0.2M; (c) 0.3M; (d) 0.5M.*

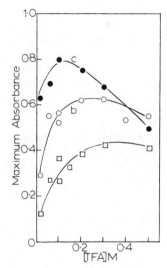

*Figure 9.   Variation of maximum absorbances reached at (a) 590 nm, (b) 660 nm, and (c) 730 nm with trifluoroacetic acid concentrations for dichloromethane solutions of chemically degraded PVC (38)*

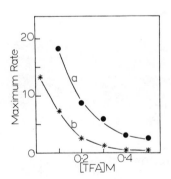

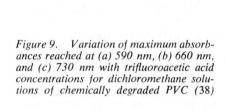

*Figure 10.   Variation of maximum rate of increase in absorbance at (a) 730 and (b) 790 nm with trifluoroacetic acid concentration of chemically degraded PVC (38)*

Samples of the chemically degraded PVC without added TFA
were extracted in DCM or CH at ice temperature for one hour.  The
resulting solutions were filtered and after dilution aliquots
were placed in spectrophotometer cells in the thermostated
cell compartment of a spectrophotometer and the spectra measured
at intervals over about 200 minutes.  The spectra of the initial
polyene solution in DCM before dilution is shown in figure 11
and successive spectra of diluted solutions in DCM and CH in
figures 12 and 13 respectively.

The striking difference between the polyene distributions
in DCM and in CH or THF and the formation of the polyene - TFA
interactions in DCM but not in CH or THF may have some
relevance to the role of HCl in the degradation of PVC.  An
examination of polyene distributions found by various workers
for PVC degraded thermally or photochemically under different
conditions shows clearly that the efficiency with which HCl is
removed as the reaction progresses is a critical factor.  The
distributions we observed in THF and in DCM seem to represent
the extremes which would result from complete removal or
complete retention respectively of evolved HCl.  The difficulty
of distinguishing whether initiation, propagation or termination
are most affected by HCl has already been mentioned and it now
appears that for systems in which the HCl is not removed, the
migration of short polyene sequences along the polymer chain may
be another complicating factor.  The formation of long sequences
by the accumulation of short ones may occur following the
establishment of a prototropic equilibrium and migration of
polyenes along the polymer chain as described in equations
13-14.

$$\sim CH_2 \cdot CHCl \cdot CH = CH - CH = CH \cdot CH_2 \cdot CHCl \sim$$

$$\text{-H}^+ \Big\uparrow \quad \Big\downarrow \text{+H}^+$$

$$\sim CH_2 \cdot CHCl \cdot CH = CH = CH = CH_2 \cdot CH_2 \cdot CHCl \sim \qquad ----(12)$$

$$\text{+H}^+ \Big\uparrow \quad \Big\downarrow \text{-H}^+$$

$$\sim CH_2 \cdot CCl = CH - CH = CH - CH_2 \cdot CH_2 \cdot CHCl \sim \qquad ----(13)$$

The overall results would be:-

$$\sim (CH = CH)_a \sim\sim (CH = CH)_b \sim \rightarrow \sim (CH = CH)_{a+b} \sim ----(14)$$

the rate of which would depend on the proximity of the two
sequences in the polymer.

The most likely lengths of the polyenylic cations
responsible for the maxima at 590, 660, 730 and 790 nm can best
be inferred by reference to the data of Sorenson (40) who has
defined an empirical relationship, $[\lambda_{max} = (330.5 + 65.5n)nm]$
for ions of type I.

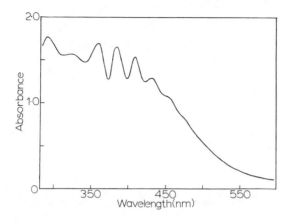

*Figure 11.   Spectrum of dichloromethane solutions of chemically degraded PVC extracted at 0.2 °C for 1 h before dilution*

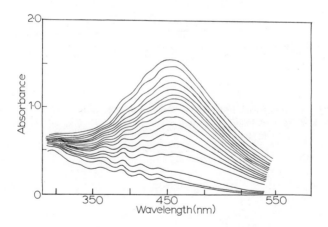

Journal of Applied Polymer Science

*Figure 12.   Changes with time in the absorption spectrum at 35 °C of dichloromethane solutions of chemically degraded PVC (38). Increasing absorbances correspond to 0, 10, 20, 40, 50, 60, 70, 80, 90, 100, 110, 120, 130, 150, 170, 200, and 230 min.*

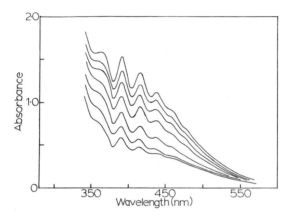

Journal of Applied Polymer Science

*Figure 13.    Changes with time in the absorption spectrum at 32°C of cyclohexa-none solutions of chemically degraded PVC (38).    Increasing absorbances corre-spond to 0, 75, 120, 190, 240, 285, and 370 min.*

$$\text{(I)}$$

The absorption maxima for individual members of the series are separated from each other by about 65nm which is also the separation between successive maxima in our systems. Precise identification is made more difficult by the fact that the exact positions of the maxima are solvent dependent and are shifted hypsochromically as the solvent polarity is decreased. In addition it is probable that each absorbance has a shoulder on the short wavelength side of the maximum. Application of Sorenson's data would associate the peaks at 590, 660, 730 and 790 nm with ions derived from pentaenes, hexaenes, heptaenes and octaenes respectively and interconversion of $A_{590}$ to $A_{660}$, $A_{660}$ to $A_{730}$ and so on corresponds to an increase in the length of the ions by one unit in each case. Wasserman (41) has also studied polyenic cations derived by protonation of various polyenes but the reliability of his spectral data has been questioned (40) due to the high reactivity of the species concerned.

The data of figures 9 and 10 which describe the effect of varying TFA concentrations on the absorbance maxima and the maximum rates of increase of absorbance enabled some semi-quantitative conclusions to be drawn. As the TFA concentration is increased, the value of the absorbance maximum reached increases to a maximum value then begins to decrease (figure 9). The values of the extinction coefficients for the various ions are not significantly different (42) and so changes in absorbance probably reflect relative concentrations. The maximum for $A_{660}$ is reached at lower TFA concentration than $A_{590}$ and it decreases faster, the maximum for $A_{730}$ at even lower TFA concentration with an even faster decrease. It seems that the lower the TFA concentration, the fewer polyenes are protonated initially and the higher the tendency for migration to occur. This explains why maximum rates of increase of absorbance due to the higher ions (730 and 790 nm) occur at low TFA concentrations and why maximum concentrations of large ions are formed.

The extreme photosensitivity of these species is similar to that found by Sorenson (40) for analogous but non polymeric species in strongly acid media. In those cases the principal photochemical reaction taking place was 1,5-cyclization with exclusive formation of five membered rings even though the formation of seven and nine membered rings was possible. The rates were controlled to a large extent by steric factors operating in the ions. These steric factors may be even more critical in our systems where the ions are relatively short segments in long polymer chains. Interpolymer reactions of the type shown in equation 15 which lead to cross links may also be important.

$$\sim CH_2 \cdot CHCl \cdot \overset{+}{CH} \cdot CH_2 \sim \qquad \sim CH_2 \cdot CHClCH \cdot CH_2 \sim$$
$$\sim CH = CH \sim \qquad \longrightarrow \qquad \sim \overset{+}{CH} - \overset{|}{CH} \sim \qquad ----(15)$$

These results emphasise the important role played by HCl not only as a catalyst for the dehydrochlorination process but in influencing the distribution of polyene sequences which result from the primary part of the degradation process and the photochemical cross-linking reactions of the polyenylic cations.

Evidence has accumulated from various sources which supports the idea that polyenylic cations are implicated in other aspects of PVC degradation. Molecular orbital calculations carried out by Starnes (43) show that the charge is better stabilized at the center of the formal delocalized length of the ion than at the end and that since the process represented by equation 16 becomes more favorable with increasing sequence length, this may provide an explanation for the relatively short sequence lengths.

$$----(16)$$

The possible importance of side reactions such as Friedel Craft alkylation, inter or intramolecular Diels Alder cyclization or re-addition of HCl in this context have also been emphasised (43).

The $BCl_3$ catalysed synthesis of novel block co-polymers of PVC with isobutylene is thought to involve microcations formed by extraction of a chloride ion activated by the chemical introduction of some unsaturation (44) (equation 17).

$$\sim CH_2 - (CH = CH)_n \sim CHCl \sim \qquad \longrightarrow \qquad \sim CH_2 - (CH = CH)_n - \overset{+}{CH} \sim \quad ---(17)$$
$$BCl_3 \qquad\qquad\qquad\qquad\qquad\qquad\qquad\qquad BCl_4^-$$

They have also been implicated (45) in the HCl catalysed allyl activated reinitiation of thermal degradation as well as in the "Schlimper" type complexes which have been thought to contribute to the color of degraded PVC (46).

## References

1. E.J.Arlman, J.Polymer Sci., 12, 543, (1954).
2. D.Druesedow and C.F.Gibb, Nat.Bur.Std.Circ. 626, 69 (1953).
3. A.Riecke, A.Grimm and H.Mucke, Kunstoffe, 52, 265 (1962).
4. G.Talamini, G.Cinque and G.Palma, Materie Plastiche, 30, 317 (1964).
5. A.Crosato Arnaldi, G.Palma and G.Talamini, Materie Plastiche, 32, 50 (1966).
6. D.E.Winkler, J.Polym.Sci., 35, 3 (1959).
7. D.Braun and R.F.Bender, Eur.Pol.J.,Supp., 269 (1969).
8. S.Van der Ven and W.F. de Wit, Angew Mak.Chem., 8, 143 (1969).
9. M.Thallmeier and D.Braun, Makromol Chem., 108, 241 (1967).
10. W.C.Geddes, Rubber Chem. and Technol., 40, 178 (1967).
11. M.Onozuka and M.Asahira, J.Macromol.Sci., Rev.Macromol. Chem., C3, 235 (1969).
12. Z.Mayer, J.Macromol.Sci., Revs.Macromol Chem., C10(2), 263 (1974).
13. V.P.Gupta and L.E. St. Pierre, J.Polymer Sci., A-1, 11, 1841 (1973).
14. B.Dodson and I.C.McNeill, J.Polymer Sci., A-1, 14, 353 (1976).
15. R.A.Papko and V.S.Pudov, Polymer Sci., USSR, 16, 1636 (1974).
16. K.P.Nolan and J.S.Shapiro, (a) J.Chem.Soc., Chem.Comm., 490 (1975), (b) J.Polymer Sci., Symposium No.55, 201 (1976).
17. T.Hjertberg and E.M.Sorvik, J.Appl.Polymer Sci., 22, 2415 (1978).
18. K.S.Minsker, V.P.Malinskaya, M.I.Artsis, S.D.Razumovskii and G.E.Zaikov, Dokl.Akad.Nauk., USSR, 223, 138 (1975).
19. M.I.Abdullin, V.P.Malinskaya, S.V.Kolesov and K.S.Minsker, Second International Symposium on PVC, Lyon-Villeurbanne, France, 273 (1976).
20. Von Yu. Moiseev, M.Artsis, K.Minsker and G.Zaikov, Kunststoff, 2, 39 (1976).
21. T.Morikawa, Chem.High Polym. (Japan), 25, 505 (1968).
22. G.A.Rasuvaev, L.S.Trotskaya and B.B.Troitskii, J.Polymer Sci., A-1, 9, 2673 (1971).
23. J.Millan, E.L.Madruga and G.Martinez, Angew.Makromol. Chem., 45, 177 (1975).
24. A.A.Berlin, G.A.Vinogradov and V.M.Kobryanskii, Izu.Akad. Nauk., SSSR, Ser. Khim., 1192 (1970).
25. Y.Pocker, K.D.Stevens and J.J.Champoux, J.Amer.Chem. Soc., 91, 4199, 4205 (1969).
26. M.M.Zofar and R.Mahmood, Europ.Polymer J., 12, 333 (1976).
27. E.D.Owen and J.I.Williams, J.Polymer Sci., Pol.Chem.Ed., 12, 1933 (1974).
28. L.M.Quick and E.Whittle, Can.J.Chem., 45, 1902 (1967).
29. W.H.Gibb and J.R.MacCallum, Europ.Pol.J., 10, 533 (1974).

30.  R.J.Bailey, Ph.D. thesis, University of Wales, 1972.
31.  M.Imomoto and T.Nakaya, Kogyo Kagaku Zasski, 68, 2285 (1965).
32.  G.C.Marks, J.L.Benton and C.M.Thomas, SCI Monograph 26, 204 (1967).
33.  D.Braun, Pure and App.Chem., 26, 173 (1971).
34.  W.I. Bengough and I.K. Varma, Europ.Polymer J., 2, 61 (1966).
35.  W.I.Bengough and G.F.Grant, Eur.Polymer J., 4, 521 (1968).
36.  L.V.Smirnov and V.I.Grachev, Vyoskomol Svedin., A14, 335 (1972).
37.  G.Palma and M.Carenza, J.Appl.Polymer Sci., 16, 2485 (1972).
38.  E.D.Owen, I.Pasha and F.Moayeddi, J.Appl.Polymer Sci., (in the press).
39.  Y.Shindo, B.E.Read and R.S.Stein,Makromol.Chem., 118, 272 (1968).
40.  T.S.Sorenson, In "Carbonium Ions," Vol. 11, G.A.Olah and P. von R.Schleyer, (Eds.), Interscience, N.Y., 1970, p.807.
41.  A.Wasserman, J.Chem.Soc., 4329 (1965); 979, 983, 986 (1959).
42.  G.A.Olah, C.U.Pittman Jnr., and M.C.R.Symons, in "Carbonium Ions," Vol. 1, G.A.Olah and P. von R.Schleyer, (Eds.), Interscience, N.Y., 1968, p.153.
43.  R.C.Haddon and W.H.Starnes, Adv.Chem.Ser., 169 (Stab. Deg.Polym.), 333 (1978).
44.  S.N.Gupta and J.P.Kennedy, Polymer Bulletin, 1, 253 (1979).
45.  T.T.Nagy, T.Kelen, B.Turcsanyl and F.Tudos, Polymer Bulletin, 2, 77 (1980).
46.  R.Schlimper, Plaste Kautschuk, 13, 196 (1966).
47.  A.R.Amer and J.S.Shapiro, J.Macromol.Sci.-Chem., A14(2), 185, (1980).
48.  C.Decker and M.Balandier, Proc. 26th IUPAC Int.Symp. Macromol., Mainz, W.Germany (1979), 588.
49.  J.Verdu, J.Macromol.Sci.-Chem., A12(4), 551 (1978).

RECEIVED October 27, 1980.

# Studies in Photophysical and Photodegradation Processes in Aromatic Polyester Yarns

P. S. R. CHEUNG[1], JACK A. DELLINGER[2], W. C. STUCKEY[3], and C. W. ROBERTS

Textile Department, Clemson University, Clemson, SC 29631

With the first plastic material came the first plastic degradation problem. In many cases, it was immediately clear that one of the major causative agents for degradation was radiative energy in 300-800 nm range. Adding to the cause of the problem was the role, in degradation, of certain environmental constituents: water, oxygen, atmospheric 'pollutants,' and other constituents in the plastic itself added either for a specific purpose or present as an impurity.

In terms of photodegradation mechanisms, polymers can be divided into two groups: those in which degradation is initiated by an ultraviolet absorbing 'impurity' produced during production and those which contain recurring units having high absorption in the near ultraviolet region. Poly(ethylene terephthalate), PET, belongs to the second group and has only in recent years become the subject of systematic studies regarding light stability.

Day and Wiles (1-6) have reported the most complete investigation pertaining to the photochemical aspects of PET degradation. Merrill and Roberts (7) have reported the effects of Disperse Red 59 on the photodegradation of PET. Included in this study are the effects of such variables as $TiO_2$ and water. Cheung and Roberts (8) have reported the photophysics and photodegradation of a series of model terephthalate esters in relation to PET degradation.

PET in comparison to other polymers shows a reasonable degree of lightfastness. Under normal conditions, degradation, resulting in significant losses in mechanical properties, occurs only after long term exposure to terrestrial light. Therefore, little work has been reported on the photostabilization of PET.

Current Address: [1]Western Company of North America, Ft. Worth, TX 76101
[2]Akzona, Inc., Enka, NC 28728
[3]Coats & Clark, Toccoa, GA 30577

0097-6156/81/0151-0239$06.00/0
© 1981 American Chemical Society

Photostabilization is usually accomplished by adding a stabilizer that falls into one of three groups:  light screens, ultraviolet absorbers, or quenching compounds.  Quenchers need not reflect or absorb the harmful radiation but may merely act to quench the excited state of the species to be protected.  The most commonly studied method of quenching the excited state is for the quencher species to accept the excitation energy by an energy transfer process then dissipate the same energy in a harmless manner.

Photostabilizers, regardless of their mechanism of action, have been added as low molecular weight materials at some point in processing.  Subsequently, these stabilizers are often lost in further processing due to their volatility or else later migrate to the surface and evaporate.  One method which avoids this modifies the polymer to include the quencher as an additional monomer in the polymerization.  This paper will describe some recent efforts in our laboratory to pursue this latter approach in the stabilization of poly(ethylene terephthalate).

## Experimental

The poly(ethylene terephthalate-co-esters) were prepared as previously described ($\underline{9}$); after knitting, scouring, and deknitting, the yarns were examined, tested and exposed in a standard fashion ($\underline{9}$) to 3000 Å radiation.

Ultraviolet absorption spectra were obtained from a Cary 118C Spectrophotometer.  Luminescence measurements were obtained from a Perkin-Elmer Model MPF-3 Fluorescence Spectrophotometer equipped with Corrected Spectra, Phosphorescence and Front Surface Accessories.  A Tektronix Model 510N Storage Oscilloscope was used for luminescence lifetime measurements.  Fiber irradiation photolyses were carried out in a Rayonet Type RS Model RPR-208 Preparative Photochemical Reactor equipped with a MGR-100 Merry-go-Round assembly.

The light intensity of the 3000 Å lamps was determined as previously described ($\underline{9}$).  Yarn samples were knit on a Lawson Fiber Analysis Knitter (FAK), and yarn tensile testing was performed on an Instron Model 1101 (TM-M) constant rate of extension testing machine.

The model compounds were obtained from commercial sources and purified using standard recrystallization techniques from spectro-grade solvents.

## Discussion

Poly(ethylene terephthalate), (PET), is a thermoplastic polymer widely used in the production of fibers and films; on exposure to near ultraviolet light, PET fibers tend to lose their elasticity and break easily; PET films become discolored, brittle and develop crazed surfaces.  Such deterioration in properties has been attributed to photochemical reactions initiated by the

absorption of near ultraviolet radiation.  PET has been shown to
absorb strongly below 315 nm but shows no absorption at wavelengths greater than 320 nm (10).  The degradation and subsequent changes of PET have been studied as a function of many
variables, such as irradiation wavelength, irradiation atmosphere,
irradiation time, and polymer additives (2-4, 10-16).

Photophysical Processes in PET and Model Compounds.  The
photophysical processes in many polymer, copolymer, and polymer-
additive mixtures have been studied (17, 18, 19).  However, until
recently, few investigations have been made concerning the photophysical processes available to the aromatic esters in either
monomeric or polymeric form.

We (8) have reported the photophysical processes of a series
of model esters of PET, and tentatively assigned the fluorescence
and phosphorescence of the aromatic esters as $^1(n, \pi^*)$ transitions, respectively.  We (9) also performed an extensive study
of the photophysical processes available to dimethyl terephthalate
(DMT) in order to relate this monomeric species to the PET polymer.  In 1,1,1,3,3,3-hexafluoro-2-propanol (HFIP) (Table I), DMT
has three major absorptions which are according to Platt's nota-
tion:  191 nm, $^1A \rightarrow ^1B$, $\varepsilon = 40,620$ 1 mole$^{-1}$ cm$^{-1}$; 244 nm, $^1A \rightarrow ^1L_a$,
$\varepsilon = 23,880$ 1 mole$^{-1}$ cm$^{-1}$; 289 nm, $^1A \rightarrow ^1L_b$, $\varepsilon = 1780$ 1 mole$^{-1}$ cm$^{-1}$.
In a less polar solvent, 95% ethanol, absorptions at 241 and 286
nm were reported with the $^1A \rightarrow ^1B$ transition being obscured by solvent absorption.  Also in HFIP, DMT displays a fluorescence approximately 100 times as intense as in ethanol solution was
reported.

TABLE I.  Absorption Characteristics of Dimethyl Terephthalate

| Solvent | $\lambda$(nm) | $\varepsilon$(1 cm$^{-1}$ mole$^{-1}$) |
|---|---|---|
| Hexafluoroisopropanol | 191.0 | 40,620 |
| | 244.0 | 23,880 |
| | 289.0 | 1,780 |
| | 297.0 (s)* | 1,379 |
| 95% Ethanol | 241.2 | 20,630 |
| | 286.2 | 2,074 |
| | 294.0 (s)* | 1,298 |

*Shoulder

Luminescence studies of DMT in an ethanol glass at 77°K are
reported to have shown a highly structured phosphorescent emission with maxima at 391, 404, 418, 432, and 446 nm and a mean

lifetime ($\tau$) of 2.2 seconds (Figure 1). This structured phosphorescent emission has been postulated as derived from a $^3L_a(\pi,\pi^*)$ state. We assigned the following electronic state energies for DMT: $^1S_1 \sim 33,000 cm^{-1}$, $^1S_2 \sim 42,000 cm^{-1}$ and $^3T_1 \sim 26,000 cm^{-1}$. The absorption spectra of Mylar$^R$ films have been reported by several authors (10,15,20). All agree that even the thinnest, commercially available films are transparent to wavelengths greater than 320 nm but absorb at wavelengths less than 314 nm strongly. Marcotte et al. (15) and Takai et al. (20) have reported in a very thin film ($\sim$500 Å), a strong absorption maximum exists at about 240 nm and a weaker absorption maximum exists at about 290 nm. We have reported the absorption properties of dilute solutions of PET dissolved in HFIP and assigned the following absorptions: 191 nm, 245 nm, and 291 nm corresponding to the $^1A \rightarrow ^1B$, $^1A \rightarrow ^1L_a$, and $^1L_a \rightarrow ^1L_b$ transitions observed in DMT (9). The room temperature luminescence of dilute PET solution using HFIP as a solvent showed an excitation spectrum having maxima at 255 and 292 nm and an emission spectrum consisting of a single structureless band centered at 325 nm. The emission maximum is independent of excitation wavelengths. Hence, it has been shown that in dilute solution the room temperature absorption and emission properties of DMT and PET are nearly identical.

However, the case of luminescence of PET fibers and films is not so easily interpreted and has recently been the subject of several studies (2,7,9,21,22,23). There is general agreement that PET does have a luminescent state and that the observed emission is not merely an impurity. The origin of the fluorescence has remained the subject of debate for the past decade.

Phillips and Schug (24) have suggested that the 390 nm emission, observed when PET is excited with high energy electrons, might be from a triplet state or an excimer. Since the triplet states of both PET and DMT are lower in energy ($\sim$450 nm), it is unlikely that the emission is from a triplet state. In addition, excimer formation and emission should not effect the absorption-excitation processes; therefore, it is unlikely that the 390 nm emission is from an excimer.

Merrill and Roberts (7) have examined both PET films and fibers and have attributed the fluorescence (excitation 342 nm, emision 388 nm) to a $^1(n,\pi^*)$ transition. They have proposed a $^1(n,\pi^*)$ transition, since the observed fluorescence is at lower energy than the observed phosphorescence (excitation 313 nm, emission 452 nm, 1.2 sec), which they have proposed from a $^3(\pi,\pi^*)$ state.

Subsequently, we reported the same experimental data as Merrill and Roberts but we have attributed the fluorescence to a $^1(\pi,\pi^*)$ transition rather than a $^1(n,\pi^*)$ transition. We also pointed out that the $^1(n,\pi^*)$ state of PET is probably at higher energy than the $^1(\pi,\pi^*)$ state (Figure 2). We attributed the red shift in the fluorescence excitation and emission to aggregates of monomeric units fixed in specific geometry in the polymer matrix.

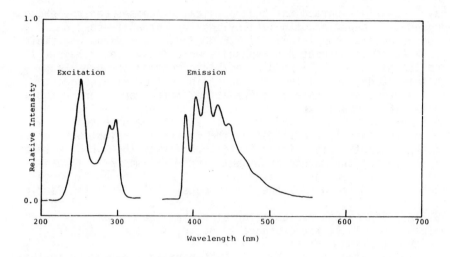

Journal of Applied Polymer Science

*Figure 1. Uncorrected phosphorescence excitation and emission spectra of dimethyl terephthalate (5 × 10⁻⁵M) in 95% ethanol at 77 K, excitation scan: Em λ 418 nm; emission scan: Ex λ 250 nm; lifetime (τ) 2.2 sec (9)*

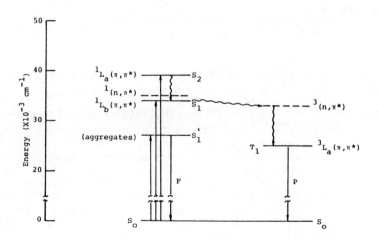

Journal of Applied Polymer Science

*Figure 2. Electronic energy level diagram and transitions for poly(ethylene terephthalate); (— — —) estimated levels (9)*

Photophysical Processes and Photodegradation of Poly(ethylene terephthalate-co-2,6-naphthalenedicarboxylate) Copolymers. We have recently reported the photophysical processes and the photo-degradative behavior of poly(ethylene terephthalate-co-2,6-naphthalenedicarboxylate), PET-2,6-ND, copolymer yarns containing 0.5 - 4.0 mole percent 2,6-naphthalenedicarboxylate, 2,6-ND (9) and the parent naphthalenedicarboxylate monomer, Figure 3 and 4. In this study, we related the change in photophysical processes to the decrease in rate of degradation of the copolymers relative to the PET homopolymer.  Copolymers containing as little as 0.5 mole percent 2,6-ND showed that over 75% of the observed phosphorescence is sensitized by the transfer of electronic energy of the terephthalate triplet state to activate the 2,6-ND triplet state.  Using the method of Inokuti and Hirayama (27), it was shown that the energy transfer process proceeds by an exchange mechanism and approaches the Perrin model (28).  From the Perrin model we calculated a critical transfer radius, $R_0$, equal to 19.7 Å.  Since this radius is greater than the normally expected values (10 - 15 Å) for exchange to occur, we concluded that energy migration must also be operating.

In addition to the change in photophysical processes in the copolymer system, we found a significant decrease in rate of photodegradation of the copolymers.  Assuming a zero order rate of photodegradation for the initial stages, the calculated rates of $2.0 \times 10^{-19}$ and $0.7 \times 10^{-19}$ percent breaking strength loss/quantum absorbed/$cm^2$ for PET homopolymer and the copolymer yarns containing 4.0 mole percent 2,6-ND were found respectively.

Combining the luminescence data and the degradation data, we concluded that most photodegradation reactions of PET, as expected, proceed via the triplet state; it was further concluded that the quenching of the terephthalate triplet state, through triplet-triplet energy transfer, by the 2,6-ND, effectively interrupts the photodegradation sequence, thus stabilizing the polymer.

Photophysical Processes in Dimethyl 4,4'-Biphenyldicarboxy-late (4,4'-BPDC).  The ultraviolet absorption spectrum of dimethyl 4,4'-biphenyldicarboxylate was examined in both HFIP and 95% ethanol.  In each case two distinct absorption maxima were recorded, an intense absorption near 200 nm and a slightly less intense absorption near 280 nm.  The corrected fluorescence excitation and emission spectra of 4,4'-BPDC in HFIP at 298°K shows a single broad excitation band centered at 280 nm with a corresponding broad structureless emission band centered at 340 nm. At 77°K, the uncorrected phosphorescence spectra shows a single broad structureless excitation band centered at 298 nm, and a structured emission band having maxima at 472 and 505 nm with a lifetime, $\tau$, equal to 1.2 seconds.

In 95% ethanol, 4,4'-BPDC shows a strong fluorescence and relatively weaker phosphorescence (relative intensity $\sim 10^{-2}$ times the fluorescence).  This indicates, though perhaps unexpected, the lowest energy singlet state is a $^1(\pi,\pi^*)$ state

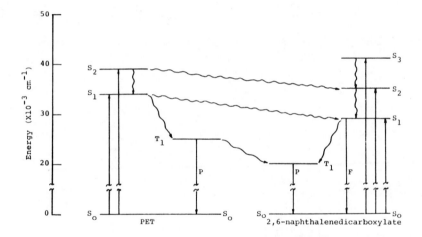

Journal of Applied Polymer Science

*Figure 3.   Electronic energy level diagram and transitions for poly(ethylene tereph-thalate-co-2,6-naphthalenedicarboxylate) yarn (9)*

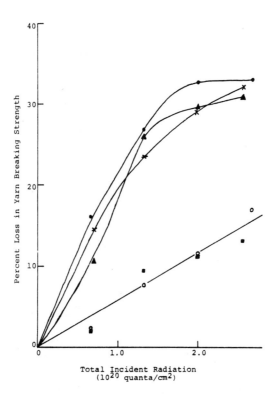

*Figure 4.    Effect of radiation on the poly(ethylene terephthalate-co-2,6-naphthalene-dicarboxylate) yarns; mole % of 2,6-DMN: (●) 0.0; (✕) 0.5; (▲) 1.0; (○) 2.0; (■) 4.0 (9)*

rather than a $^1(n,\pi^*)$ state which would produce a weak
fluorescence and a relatively intense phosphorescence. Further
evidence that the lowest energy excited singlet state for 4,4'-
BPDC is a $^1(\pi,\pi^*)$ state is shown by the slight red shift of the
fluorescence in the more polar solvent, HFIP. Polar solvents
raise the relative energy of $(n,\pi^*)$ states and many lower the
energy of $(\pi,\pi^*)$ states. From the luminescence data, the
following electronic state energies have been calculated for
4,4'-BPDC: $^1L_a(\pi,\pi^*){\sim}35{,}500$ cm$^{-1}$, $^1L_b(\pi,\pi^*){\sim}32{,}000$ cm$^{-1}$,
$^3L_a(\pi,\pi^*){\sim}22{,}000$ cm$^{-1}$ (Figure 5).

  Energy Transfer Studies with Dimethyl Terephthalate (DMT)
and 4,4'-BPDC. Several attempts were made to determine if
energy transfer could occur from an excited DMT molecule to a
4,4'-BPDC molecule in a rigid ethanol glass at 77°K. These
studies were accomplished by adding various amounts (20 - 50 mole
percent) 4,4'-BPDC to a known concentration (5.0 x 10$^{-4}$ M) of DMT.
The change in emission intensity at 418 nm, which is exclusively
emission from DMT, was then measured with excitation at 298 nm.
  The results of this study show a definite quenching of the
418 nm phosphorescence emission of DMT. One would expect that
the quenching effect, in a rigid glass, would fit the Perrin model
(73). A plot in ln $\phi_o/\phi$ versus concentration of 4,4'-BPDC yielded
a straight line, the slope of which was identified with NV. The
radius, $R_o$, of the active volume of quenching sphere was calculated by the following equation:

$$R_o = (\frac{3V}{4\pi})^{1/3}.$$

  A least square analysis, the slope, was determined to equal
3.18 x 10$^3$ 1 mole$^{-1}$, which yielded a value of $R_o$ equal to 108 Å.
In addition, it can be shown for the concentration range of the
4,4'-BPDC used, assuming each molecule occupies a spherical
volume, the average radius of this volume is about 108 Å. This
calculation predicts, on the average, the probability of an
excited DMT molecule having a 4,4'-BPDC molecule within the
required 15 Å for energy transfer to occur by the exchange
mechanism, which would be spin allowed, is small.
  If neither mode of energy transfer is acceptable, a different
explanation of the apparent quenching of the DMT phosphorescence
must be put forth. It must be recalled that both DMT and 4,4'-
BPDC absorb 298 nm light, which introduces the argument that
competitive absorption causes the apparent quenching effect.
Since 4,4'-BPDC has no phosphorescence emission at 418 nm,
light absorbed by the 4,4'-BPDC molecules is essentially lost to
the detector monitoring the 418 nm emission. This can easily be
seen by monitoring the change in excitation spectra of the mixed
solutions as a function of 4,4'-BPDC concentration. At zero
concentration of 4,4'-BPDC and 5.0 x 10$^{-4}$ M DMT concentration,
since the rotation of the chopper limits the amount of light
reaching the sample, most of the available light is absorbed by
the DMT and the excitation maxima at 250 and 298 nm reflect the

relative intensity of the xenon source at these respective wave-lengths. Since the output of the source is greater at 298 nm than the maximum excitation appears at 298 nm. As the 4,4'-BPDC concentration is increased and competes for absorption of the 298 nm radiation, the excitation scan of 418 nm emission reflects the probability of DMT excitation rather than source intensity. Thus, the 250 nm band, corresponding to the more probable $^1A \rightarrow {}^1L_a$ transition, becomes the predominate band.

The conclusion from the monomer solvent studies is that, in nearly equal molar solutions, DMT and 4,4'-BPDC compete for absorption of the 298 nm radiation. However, the results also show that, even in equal concentrations, the DMT emission, when excited by 298 nm light, is several times as intense as the 4,4'-BPDC emission at 472 nm. It must be emphasized that these studies do not preclude the existence of energy transfer from excited DMT to 4,4'-BPDC. From the volume calculation used above, it can be shown that a concentration of $\sim$ 0.1 M 4,4'-BPDC is needed to assume an occupied volume with radius of 15 Å, the required distance for the exchange mechanism.

Photophysical Processes in Poly(ethylene terephthalate-co-4,4'-biphenyldicarboxylate) (PET-co-4,4'-BPDC). The absorption and luminescence properties of PET are summarized above. At room temperature the absorption spectrum of PET-co-4,4'-BPDC copolymers, with concentrations of 4,4'-BPDC ranging from 0.5 - 5.0 mole percent, showed UV absorption spectra similar to that of PET in HFIP. The corrected fluorescence spectra of the copolymers in HFIP exhibited excitation maxima at 255 and 290 nm. The emission spectrum displayed emission from the terephthalate portion of the polymer, when excited by 255 nm radiation, and emission from the 4,4'-biphenyldicarboxylate portion of the polymer when excited with 290 nm radiation.

Examination of the corrected room temperature fluorescence properties of PET yarns revealed an excitation maximum at 342 nm with a corresponding emission maximum at 388 nm. At 77°K, in the uncorrected mode, the fluorescence spectra of PET yarns exhibited a structured excitation having maxima at 342 and 360 nm and a shoulder at 320 nm. At 77°K, PET yarns displayed a structured emission with maxima at 368 and 388 nm. As in solution, the copolymer yarns showed both fluorescence from the terephthalate portion of the polymer and the 4,4'-biphenyldicar-boxylate portion of the polymer. Excitation at 342 nm produced an emission band centered at 388 nm. This excitation and emission correspond to the PET homopolymer emission. Excitation with about 325 nm light produced an emission with a maximum near 348 nm from the 4,4'-biphenyldicarboxylate portions of the polymer.

The uncorrected phosphorescence excitation and emission spectra of PET yarn at 77°K show an excitation maximum at 310 nm and emission at 452 nm with a lifetime, $\tau$, equal to 1.2 seconds

(Figure 6). The phosphorescence spectra of the copolymer yarns showed excitation in the 305 - 310 nm range with corresponding emission maxima at 480 and about 515 nm, corresponding lifetimes equal 1.2 seconds. In the copolymer yarns containing 0.5 - 2.0 mole percent 4,4'-BPDC a small shoulder was observed at 452 corresponding to the PET homopolymer phosphorescence.

The bands at 193.0, 245.5, and 289.5 nm in the absorption spectrum of PET in HFIP have been assigned as the $^1A \rightarrow {}^1B$, $^1A \rightarrow {}^1L_a$, and $^1A \rightarrow {}^1L_b$ transitions of DMT, respectively. These bands predominate in the absorption spectra of the copolymers. As the concentration of 4,4'-BPDC increases, an increase in the intensity of the band at 289.5 nm was observed. This is the result in the increased intensity of the $^1A \rightarrow {}^1L_a$ transition of the 4,4'-BPDC in this region. In dilute HFIP solutions the copolymers show a fluorescent emission in the 326 - 338 nm range when excited with 255 nm radiation. This emission corresponds to emission from the terephthalate units of the copolymer. However, excitation with 290 nm radiation produces an emission that is red shifted relative to the terephthalate emission. This emission originates from the 4,4'-biphenyldicarboxylate units in the copolymer. The relative increase in 4,4'-BPDC emission intensity can be observed; in the copolymer containing 0.5 mole percent 4,4'-BPDC, a slight red shift in the terephthalate emission was observed as a result of the presence of the longer wavelength emission of the 4,4'-BPDC. Similarly, the emission of the 4,4'-biphenyldicarboxylate was blue shifted from the expected 340 nm due to the presence of emission from the terephthalate unit. At 4.0 mole percent 4,4'-BPDC in the copolymer the 4,4'-biphenyldicarboxylate emission was observed at the expected 340 nm wavelength; this fluorescence emission intensity of the 4,4'-BPDC unit is greater than the fluorescence emission of the terephthalate unit of the copolymer.

In the yarns, the fluorescence of the 4,4'-biphenyldicarboxylate unit is distinct and predominate both at 298 and 77°K. Examination of the phosphorescence spectra of the PET and PET-co-4,4'-BPDC yarns revealed three emission maxima. In the PET homopolymer excitation with 310 nm radiation produced an emission at 452 nm from the terephthalate chromophore. In the copolymers excitation with either 305 or 310 nm radiation produced emission spectra with distinct maxima at 480 and $\sim$ 515 nm ($\tau$ 1.2 sec), and a shoulder near 452 nm ($\tau$ = 1.2 sec). The maxima in the phosphorescence spectra were assigned as emission from the 4,4'-biphenyldicarboxylate units of the copolymer. The observed emissions are bathochromatically shifted from the emission of 4,4'-BPDC in a glassed solvent. This is supported by the observed emissions from solid 4,4'-BPDC at 520 and 560 nm ($\tau$ = .3 sec) when excited with 340 or 356 nm radiation.

The phosphorescence characteristics of the copolymer yarns are somewhat unexpected. It has been shown previously that the terephthalate emission should predominate in the phosphorescence spectra. It has also been shown previously that in mixed

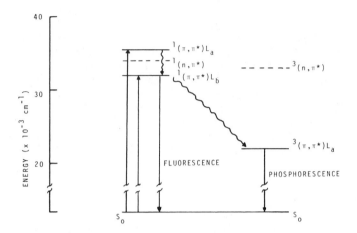

Journal of Applied Polymer Science

*Figure 5. Electronic energy level diagram and transitions for dimethyl 4,4'-biphenyldicarboxylate; (– – –) estimated levels based on dimethyl terephthalate (33)*

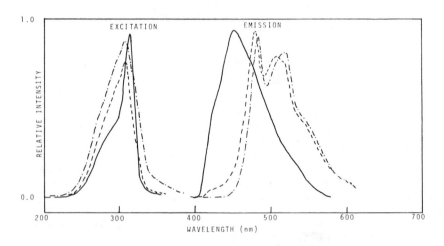

Journal of Applied Polymer Science

*Figure 6. Uncorrected phosphorescence spectra of poly(ethylene terephthalate) (———) and poly(ethylene terephthalate-co-4,4'-biphenyldicarboxylate) containing 0.5 mol % (– – –) and 2.0 mol % (– · – ·) 4,4'-BPDC; excitation scans: Em λ (———) 452 nm, (– – –) 480 nm, (– · – ·) 480 nm; emission scans: Ex λ (———) 310 nm, (– – –) 305 nm. (– · – ·) 305 nm. (33)*

solutions with equimolar concentration of DMT and 4,4'-BPDC, the
terephthalate phosphorescence emission is several times as
intense as the phosphorescence emission from 4,4'-BPDC.  On this
basis one would predict that with terephthalate concentrations
at least 24 times that of the 4,4'-biphenyldicarboxylate the
terephthalate emission should predominate in the phosphorescence
spectra of the copolymer yarns.  However, in the copolymer yarn
containing only 0.5 mole percent 4,4'-BPDC phosphorescence
emission from the 4,4'-biphenyldicarboxylate units predominate,
and at 4.0 mole percent 4,4'-BPDC the terephthalate emission is
completely quenched.

The observed luminescence properties of the copolymer yarns
can be easily explained if an energy transfer mechanism is
assumed to be operating (Figure 7).  Triplet-triplet energy
transfer from the terephthalate units to the 4,4'-biphenyl-
dicarboxylate units explains both the dual fluorescent/phospho-
rescent emissions from the 4,4'-biphenyldicarboxylate units as
well as the quenched phosphorescence from the terephthalate units.

Triplet-triplet energy transfer is spin forbidden by the
long-range dipole-dipole radiationless transfer mechanism, but is
spin allowed for the electron exchange mechanism.  The constant
lifetime (1.2 sec) of the 452 nm emission indicates that the
quenching mechanism involved fits the Perrin model.  Fitting the
data to the Perrin equation and a subsequent least squares
analysis yielded a slope equal to 8.43 1 mole$^{-1}$.  From the slope
the volume of the quenching sphere was calculated to be 1.40 x
10$^{-20}$ cm$^3$, which gave a transfer radius, $R_0$, equal to 14.9 Å.
This value for the transfer radius is within the 15 Å required
for electron exchange to occur.

Phototendering of PET and PET-co-4,4'-BPDC Filament Yarns.
Both PET homopolymer and PET-co-4,4'-BPDC copolymers were
irradiated from 20 to 80 hours in the photolysis chamber.  In
order to account for the lamp aging, the phototendering rate
curves were plotted as percent loss tenacity versus total
quanta/cm$^2$ of exposure, rather than irradiation time.  The
phototendering rate curves for the homopolymer PET and PET-co-
4,4'-BPDC copolymers show that all the samples became weaker
and showed a decrease in percent elongation to break as total
quanta/cm$^2$ of exposure was increased (Figure 21).

Assuming a zero order rate of phototendering of the yarn
samples during the initial stage of photolysis and using a
least squares analysis gave a rate constant for phototendering of
PET homopolymer, $k_{PET}$, equal to 3.4 x 10$^{-19}$ percent breaking
strength loss/quantum exposure/cm$^2$.  Similar treatment of the
phototendering with increasing concentration of 4,4'-BPDC in the
copolymer to give the following phototendering rate constants:

$$k_{0.5} = 3.0 \times 10^{-19}, \; k_{1.0} = 2.6 \times 10^{-19}, \; k_{2.0} = 2.6 \times 10^{-19},$$

$$k_{4.0} = 2.0 \times 10^{-19} \; \text{percent breaking strength loss/quantum}$$

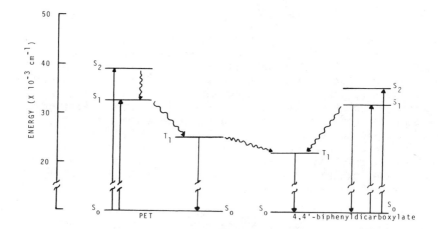

Journal of Applied Polymer Science

Figure 7.    Electronic energy level diagram and transitions for poly(ethylene tereph-
thalate-co-4,4'-biphenyldicarboxylate) yarn (33)

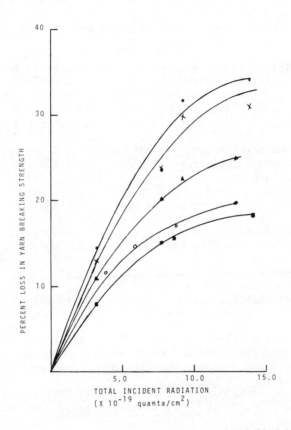

*Figure 8. Effect of radiation on the poly(ethylene terephthalate-co-4,4'biphenyl-dicarboxylate) yarns; mol % 4,4'-BPDC: (●) 0.0; (✕) 0.5; (▲) 1.0; (○) 2.0; (■) 4.0 (33)*

exposure/cm$^2$, where the subscript on the rate constant equals the mole percent 4,4'-BPDC in the copolymer.

The kinetic analyses showed that the rate of phototendering for the copolymer containing 4.0 mole percent 4,4'-BPDC decreased to about 59 percent of the rate of PET homopolymer phototendering. Tensile tests revealed after 80 hours exposure ($\sim$ 1.3 x 10$^{20}$ quanta/cm$^2$), where the rate of degradation appears to be approaching zero in all cases, the percent loss in breaking strength of the copolymer yarn containing 4.0 mole percent 4,4'-BPDC was about 54 percent of the loss incurred by the PET homopolymer.

Day and Wiles ($\underline{4}$) have shown the importance of the Norrish type II intramolecular rearrangement in PET photolysis. Based on their quantum yield measurements for formation of -COOH endgroups in both oxidative and inert environments, they report the Norrish type II rearrangement to be the predominate chain scission reaction in PET photolysis at wavelengths of 300 nm and greater. Other workers ($\underline{29, 30}$) have shown that in some systems the Norrish type II rearrangement proceeds via both the lowest excited singlet and triplet states. Doughtry ($\underline{30}$) has implied that when excitation occurs to the lower level excited states, intersystem crossing is more efficient thus allowing the triplet state to have a greater participation in the rearrangement reaction. In the case of PET photolysis, using an excitation source with a maximum output at 300 nm, the excitation process must occur to the lower level excited states. If the intersystem crossing process is efficient at this excitation, then the Norrish type II rearrangement must occur from the triplet state. This is further substantiated by a reduction in loss of tenacity with increasing concentration of triplet state quencher. The reduction in loss of tenacity may be equated with interruptions of the chain scission process(es). We conclude that the Norrish type II rearrangement in PET proceeds, for the most part, via the lowest triplet state.

Comparison of 4,4'-BPDC with Dimethyl 2,6-Naphthalene-dicarboxylate (2,6-ND) as a PET Photostabilizer. As noted before, Cheung ($\underline{9}$) reported the photostabilization of PET by copolymerization with 2,6-ND with a decrease in zero order rate of phototendering from 2.0 x 10$^{-19}$ percent breaking strength loss/quantum exposure/cm$^2$ for the PET homopolymer to 0.7 x 10$^{-19}$ percent breaking strength loss/quantum exposure/cm$^2$ at 2.0 mole percent 2,6-ND, which gives the ultimate decrease in phototendering rate.

Based on luminescence studies, we postulated triplet-triplet energy transfer by electron exchange as the mechanism of photostabilization and we calculated an active quenching sphere with a radius, $R_o$, of 19.7 Å for 2,6-ND. Because the value of $R_o$ is larger than 15 Å, we postulated that energy migration was occurring.

The 4,4'-BPDC copolymers exhibited effects similar to that reported for 2,6-ND.  Even though the absolute rates of photo-tendering differed, the change in the rate from homopolymer to 4.0 mole percent copolymer were similar.  The 2,6-ND copolymer showed a 65 percent decrease in rate compared to 59 percent for the 4,4'-BPDC copolymer.

In both cases phototendering was assumed to be a zero order reaction.  In making this assumption, if $I_0$ (intensity of radiation of exposure) is greater than $I_{abs}$ (intensity of radiation absorbed) and $I_0$ is plotted as the abscissa, then the rate of phototendering will be inversely proportional to the difference between $I_0$ and $I_{abs}$.  Since in the percent analyses, the lamps had aged giving a lower $I_0$ value, the difference between $I_0$ and $I_{abs}$ was smaller than for Cheung's work, thus yielding an apparently faster rate of phototendering.  Nevertheless, since there is only a relatively small difference in $I_0$ during the time of data collection, a zero order rate constant gives a good comparison of samples irradiated in the same time period.

The 2,6-ND monomer showed an ultimate decrease in photo-tendering rate at 2.0 mole percent, whereas the rate of photo-tendering showed a decrease from 2.0 to 4.0 mole percent 4,4'-BPDC.  Theoretically, at 1.6 mole percent quencher in PET, even in the absence of energy migration, the ultimate decrease in rate of phototendering should be achieved if the quencher mole-cules are randomly spaced thoughout the polymer matrix.  The variation in rate of phototendering for the copolymers containing 2.0 and 4.0 percent 4,4'-BPDC may simply reflect a nonrandom distribution of quencher in the fiber sample.  Since at 2.0 mole percent 2,6-ND, the copolymer exhibited an ultimate quenching effect, this may either indicate that a totally random distribution was achieved or that the observed energy migration made up for the nonrandom distribution of quencher.  If energy migration does indeed make up for the error in distribution, one must conclude that 2,6-ND is a better stabilizer than 4,4'-BPDC.

Fluorescence Analysis of Irradiated PET and PET-co-4,4'-BPDC Yarns.  The presence of a material, which emits a blue-green fluorescence, on photooxidized PET has been reported previously (2, 21).  This fluorescent material, which emits at 460 nm when excited by 342 nm energy, has been proved to be monohydroxy-tere-phthalate.

The emission spectrum of the irradiated PET yarn, when excited by 342 nm energy is totally dominated by the 460 nm emission, which has been attributed to the presence of mono-hydroxy-terephthalate, with only a shoulder as evidence of the residual fluorescence from the terephthalate units (Figure 9).  On the other hand, the exposed copolymer yarn containing 4.0 mole percent 4,4'-BPDC still exhibits the normal terephthalate fluorescence (388 nm emission) as the major band in the emission spectrum when excited with 342 nm energy.

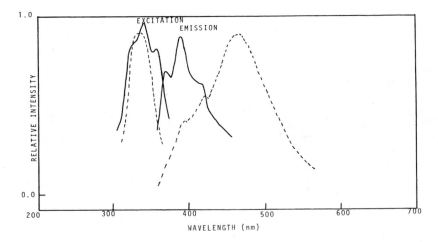

*Figure 9. Uncorrected fluorescence spectra of poly(ethylene terephthalate yarn;*
*(———) 0-h exposure; excitation scan: Em λ 388 nm; emission scan: Ex λ 342 nm;*
*(– – –) 100-h exposure to RUL 3000-Å lamps; excitation scan: Em λ 460 nm;*
*emission scan: Ex λ 342 nm.*

Photophysical Processes in Dibutyl 4,4'-Sulfonyldibenzoate
(4,4'-SD). The UV absorption spectra of dibutyl 4,4'-sulfonyl-
dibenzoate (4,4'-SD) in both HFIP and 95% ethanol showed similar
absorptions. The corrected excitation and emission fluorescence
spectra of 4,4'-SD in HFIP at 298°K showed a structured
excitation with band maxima at 236, 286, and 294 nm and a
structured emission exhibiting band maxima at 322, 372, and 388
nm. The uncorrected excitation and phosphorescence spectra of
4,4'-SD in a 95% ethanol glass at 77°K displayed excitation band
maxima at 268, 282, and 292 nm with strong phosphorescence
emission with band maxima at 382, 398, and 408 nm with a mean
lifetime ($\tau$) of 1.2 sec.

The 0-0 transition bands exhibited in the fluorescence and
phosphorescence spectra of 4,4'-SD gave the following electronic
state energies: $^1S_1 \sim 33{,}000$ cm$^{-1}$, $^1S_2 \sim 42{,}000$, and $^3T_1 \sim$
26,000 cm$^{-1}$. These energies are identical to those found in DMT
(9) which lends support to the postulate that there is little or
no through conjugation across the sulfone group.

Photophysical Processes in PET-4,4'-SD Copolymers. PET-
4,4'-SD copolymers have UV absorption spectra similar to that of
PET homopolymer in HFIP solution. Band maxima were exhibited at
about 290, 245, and 191 nm in all the polymers.

The corrected excitation and fluorescence spectra of PET-
4,4'-SD copolymers in HFIP solution or as yarns were identical
to the spectra of PET homopolymer. The uncorrected phosphore-
sence excitation and emission spectra of PET-4,4'-SD copolymer
yarns were also identical to that of the PET homopolymer yarn.
An excitation band maximum was found at 312 nm with a broad,
structureless band centered at 452 nm found in the emission
spectra. The phosphorescence mean lifetime ($\tau$) was found to be
1.2 sec.

Since the solution and yarn fluorescence and the phosphorescence spectra give no indication of the presence of the
comonomer, 4,4'-SD, we conclude that the comonomer excitation
and emission are hidden under the strong excitation and emission
bands of the dominant PET absorbing species and that a co-
absorption, co-emission process is probably occurring.

Microanalysis of the three PET-4,4'-SD copolymer yarns for
sulfur yielded concentrations in agreement with the theoretical
values. Since the 4,4'-SD comonomer was definitely incorporated
into the three copolymer yarns, the absorption and luminescence
characteristics of the copolymers point towards a co-absorption
process between 4,4'-SD and PET rather than an electronic energy
transfer process.

Phototendering of PET and PET-4,4'-SD Filament Yarns.
Samples of PET-4,4'-SD copolymer filament yarns containing 0.5 -
2.0 mole percent of 4,4'-SD as "comer" along with PET homopolymer
filament yarns were irradiated from 20 - 80 hours. Actinometry
measurements monitored the loss in radiation intensity due to
aging.

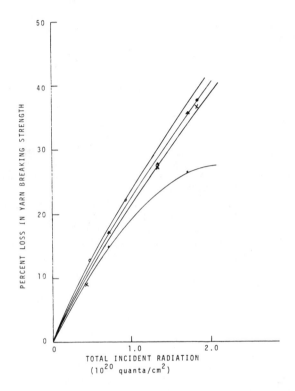

Journal of Applied Polymer Science

*Figure 10.   Effect of radiation on the poly(ethylene terephthalate-co-4,4'sulfonyl-dibenzoate) yarns: mol % of 4,4'-SD: ( · ) 0.0; (×) 0.5; (▲) 1.0; (○) 2.0 (34)*

The phototendering rate curves for the PET show that the samples become <u>weaker</u> with an increase in irradiation time; they become more <u>brittle</u> and have a lower elongation at break with increasing irradiation time (Figure 10). The copolymer yarns degrade more quickly than does the PET homopolymer yarns. Increasing concentration of the 4,4'-SD comonomer increases the rate of yarn phototendering.

<u>Fluorescence Analysis of Irradiated PET and PET-4,4'-SD Yarns.</u> As we noted above, the fluorescence emission at 460 nm in irradiated PET polymer has been attributed to the hydroxyterephthaloyl component (<u>9</u>). The fluorescence spectra of irradiated (100 hours) PET homopolymer yarns and PET-4,4'-SD copolymer yarns are identical and agree with that obtained by Day and Wiles for PET film (<u>2</u>).

A study of the relative fluorescence intensities at 460 nm of PET and PET-4,4'-SD yarns after receiving identical irradiation intensities reveals an increase in the formation of the hydroxyterephthaloyl moiety with increasing amounts of 4,4'-SD. This indicates that a photooxidative mechanism involving the second monomer may be an explanation of the increasing degradation rates.

Several studies have been performed on the photodecomposition of diaryl sulfones and polysulfones; Khodair, et. al., (<u>21</u>) demonstrated that the photodecomposition of diaryl sulfones proceeds by a free-radical mechanism with initial carbon-sulfur bond cleavage. This gives an aryl radical and an aromatic sulfonyl radical. The latter radical can react with oxygen and a hydrogen donor to eventually form the hydroxyl radical. The hydroxy radical may attack the aromatic nucleus in PET and forms the hydroxyterephthaloyl radical.

Figure 11 suggests a photooxidative mechanism which may account for the increased formation of hydroxyterephthaloyl with the presence of 4,4'-SD (<u>32</u>).

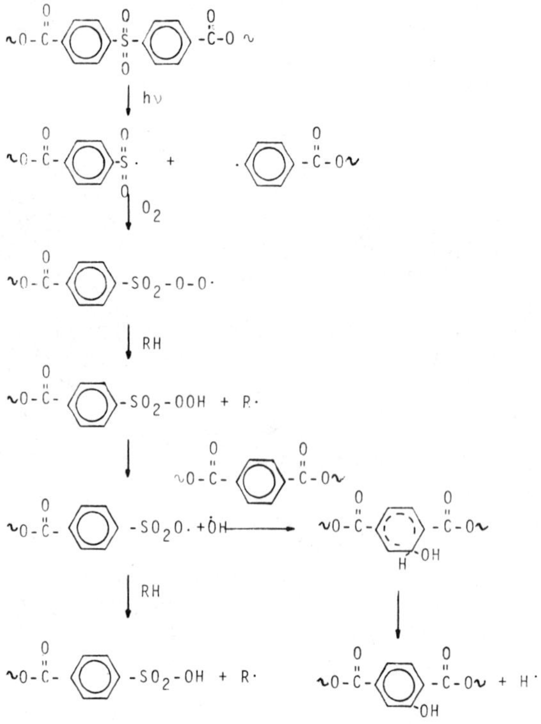

Figure 11.   Possible photooxidative mechanism occurring in poly(ethylene tereph-
thalate-co-4,4'-sulfonyldibenzoate) yarns (34)

## Literature Cited

1.  Day, M. and Wiles, D. M., Polym. Lett., 9, 665 (1971).

2.  Day, M. and Wiles, D. M., J. Appl. Polym. Sci., 16, 175 (1972).

3.  Day, M. and Wiles, D. M., J. Appl. Polym. Sci., 16, 191 (1972).

4.  Day, M. and Wiles, D. M., J. Appl. Polym. Sci., 16, 203 (1972).

5.  Day, M. and Wiles, D. M., Can. J. Chem., 49, 2916 (1971).

6.  Blais, P., Day, M. and Wiles, D. M., J. Appl. Polym. Sci., 17, 1895 (1973).

7.  Merrill, R. G. and Roberts, C. W., J. Appl. Polym. Sci., 21, 2745 (1977).

8.  Cheung, P. S. R., Master's Thesis, Clemson University, December 1974.

9.  Cheung, P. S. R., Roberts, C. W. and Wagener, K. B., J. Appl. Polym. Sci., 24, 1809 (1979).

10. Osburn, K. R., J. Polym. Sci., 38, 357 (1959).

11. Stephenson, C. V., Moses, B. C. and Wilcox, W. S., J. Polym. Sci., 55, 451 (1961).

12. Stephenson, C. V., Moses, B. C., Burks, R. C., Coburn, W. C. and Wilcox, W. S., J. Polym. Sci., 55, 465 (1961).

13. Stephenson, C. V., Lacey, J. C. and Wilcox, W. S., J. Polym. Sci., 55, 477 (1961).

14. Stephenson, C. V. and Wilcox, W. S., J. Polym. Sci., A-1, 1, 2741 (1963).

15. Marcotte, F. B., Campbell, D., Cleveland, J. A. and Turner, D. T., J. Polym. Sci., A-1, 5, 481 (1967).

16. Valk, G., Kehren, M. L. and Daamen, I., Angew, Makromol. Chem., 13, 97 (1970).

17. Birks, J. B., Photophysics of Aromatic Molecules, Wiley-Interscience, London, 1970.

18. Fox, R. B., Pure and Appl. Chem., 30, 87 (1972).

19. Ranby, B. and Rabek, J. F., Photodegradation, Photo-oxidation and Photostabilization of Polymers, Wiley-Interscience, London (1975).

20. Takai, Y., Osawa, T., Mizutani, T. and Ieda, M., J. Polym. Sci., Polym. Phys. Ed., 15, 945 (1977).

21. Pacifici, J. G. and Straley, J. M., Polym. Lett., 7, 7 (1969).

22. Yoshiaki, Takai, Mizutani, T. and Ieda, M., Jap. J. Appl. Phys., 17, 651 (1978).

23. Allen, N. S., Homer, J. and McKellar, J. F., Analyst, 101, 260 (1976).

24. Phillips, D. H. and Schug, J. C., J. Chem. Phys., 50, 3297 (1967).

25. Allen, N. S. and McKellar, J. F., Makromol. Chem., 179, 523 (1978).

26. Padhye, M. R. and Tamhane, P. S., Angew, Makromol. Chem., 69 33 (1978).

27. Perrin, J., Compt. Rend., 178, 1978 (1923).

28. Inokuti, M. and Hirayama, F., J. Chem. Phys., 43, 1978 (1965).

29. Wagner, P. J. and Hammond, G. S., J. Am. Chem. Soc., 87, 4009 (1965).

30. Doughterty, T. J., J. Am. Chem. Soc., 87, 4011 (1965).

31. Khodair, A. I., Nakabaski, T. and Karasch, N., Int. J. Sulfur Chem., 8 (1), 37 (1973).

32. Gesner, B. D. and Kelleher, P. G., J. Applied Polym. Sci., 12, 1199 (1968).

33. Dellinger, J. and Roberts, C.W., J. Applied Polym. Sci., in press.

34. Stuckey, W.C. and Roberts, C.W., J. Applied Polym. Sci., in press.

RECEIVED October 27, 1980.

# Strain-Enhanced Photodegradation of Polyethylene

DJAFER BENACHOUR and C. E. ROGERS

Department of Macromolecular Science, Case Western Reserve University, Cleveland, OH 44106

The deleterious effect of sunlight on polymeric materials has been ascribed to a complex set of reactions in which both the absorption of ultraviolet light and the presence of oxygen are participating events. The process has been termed oxidative photodegradation (in this paper, the term is used interchangeably with photooxidation), and it has been the object of many studies and review articles (1-8). These studies typically have focussed on the relationship between weathering and changes in a specific property, with mechanical and optical properties receiving considerable attention (9). However, despite intensive investigations separately on both photooxidation and deformation of polymers, very little work has been done to determine how deformation affects photodegradation. In the few studies in that area, it was found that mechanical stress accelerates polymer deterioration (10,11), but no attempt was made to establish correlations between structural changes induced by deformation and the enhanced degradation process.

In the present paper, we describe how photodegradation of low density polyethylene films was enhanced by uniaxial elongation. An explanation of the enhancement process is given based on the photooxidation and deformation mechanisms, and the photodegradation products.

Experiments done in the absence of an external stress showed that the effects of degradation crosslinking are significant at relatively short times of UV exposure, and confirmed that the photodegradation is essentially in the surface layers. The oxidized layer thickness appeared to remain more or less constant after a certain exposure.

## Experimental

### Material

Commercial low density polyethylene films were used ($\overline{M}_W \approx$ 60,000; $\rho$ = 0.920 g/ml; 55% crystallinity as measured by x-ray

0097-6156/81/0151-0263$05.00/0

diffraction; $T_m$ = 118°C; thickness = 2 mils). No attempt was made to remove additives (including any antioxidants) nor were special precautions taken to prevent air oxidation of the samples before they were exposed to UV light.

### Photooxidation Procedure

The photooxidation was carried out in a commercial weathero-meter (Q-UV, Q-panel Co., Cleveland, Ohio). This apparatus uses medium pressure mercury fluorescent UV lamps (Sunlamps F5-40, Westinghouse Electric Corp.) which emit UV light in the 273-378 nm range with a maximum intensity at 310 nm.

Films, for both mechanical and spectroscopy studies, were affixed to the specimen panels of the weatherometer. Upon com-pletion of the UV exposure, which occurred at 37°C ± 1°C in the presence of air, the films were removed and kept at room tempera-ture in the dark for at least 24 hours in order to remove any volatile oxidation products.

### Infrared Spectroscopy

The infrared spectra of the different samples were taken with a Fourier Transform infrared spectrometer (Digilab FTS-14) using the double beam mode vs. air as reference. 150 scans per sample and 100 scans per reference, at a resolution of 4 cm$^{-1}$, were taken for every sample. All spectra were stored on tape, and a digital substraction of the after- and- before UV exposure (or any other sample treatment) spectra was performed, whenever needed, by an on-line computer, thus permitting a better visuali-zation of the spectral changes in the polymer by UV- photooxida-tion.

In the case where the samples were kept elongated during UV exposure, a special film stretcher was used. This stretcher was made to fit in the FTIR sample holder, thus allowing spectra to be taken while the films are kept stretched.

### Mechanical Experiments

All tensile and stress-relaxation measurements were done using an Instron Tensile tester. The samples were cut into the dumbbell shape corresponding to the ASTM D412 type C model (total length: 4.5 in.; straight part: 1.5 in.; width: 0.25 in.). The samples were tested at a deformation rate of 1 in./min. for the simple tension experiments. In the case of stress-relaxation measurements, the samples were prestrained to 7% elongation at $\dot{\varepsilon}$ = 5 in./min. then allowed to stress relax over a 20 minute period. All mechanical testing were carried out at room tempera-ture.

## Results and Discussion

### Effects of Photooxidation on Mechanical Properties

It is well known that mechanical properties of polymeric materials are greatly deteriorated by UV exposure (2-9). The nature of this deterioration was determined using non-strained samples which were photooxidized at 37°C. Engineering stress-strain curves as a function of UV exposure are shown in Figure 1. The numbers next to each curve represent days of UV exposure. In terms of degradation, the points of interest are:

1. The large drop ($\approx$ 45%) in the stress to break between the non-oxidized sample and the oxidized ones, and

2. Both the 5- and 10-day samples fail at the same stress level ($\sigma_{nb} \approx 90 \times 10^3$ grams/cm$^2$), and differ mainly in the ultimate elongation.

Similar results were observed in the stress-relaxation experiments which are shown in Figure 2. The 5- and 10-day samples relax to the same stress level. The major difference in stress-relaxation behavior among the different samples occurs during the very beginning of the relaxation process. For that reason, and in order to better illustrate the first minutes of relaxation, the time scale is logarithmic.

We are presenting data for the 5- and 10-day samples only, but the same results were also observed for samples exposed to UV light for 6, 7, 8 and 9 days, at 37°C. These results suggest that after about 5 days of UV exposure, the applied load is supported by the same cross-section in all these samples. This means that the effective film thickness of non-degraded material does not change after 120 hours of photooxidation. Based on these findings, we are proposing the schematic model pictured in Figure 3:  the oxidation, starting at the surface, penetrates into the material bulk as the UV exposure increases. After 5 days, the extent of bulk penetration levels off, and the thickness of the oxidized layer remains effectively constant (such thickness was estimated, by taking the ratio of the load to break of the oxidized samples to that of the non-oxidized film, to be approximately 45% of the original thickness, i.e., $0.45 \times 2 = 0.9$ mils). Further oxidation will further deteriorate the already oxidized layer, resulting in the formation of more incipient cracks. This crack density increase  within the oxidized layer will, in turn, raise the probability of failure of the sample, thus resulting in a drastic decrease of the elongation to break when the films are stretched. This is shown in Figure 4. It can be seen that at relatively short UV exposures (2 to 4 days) there is a very slight increase of the ultimate elongation. This result, also observed by other workers (11,12), is attributed to oxidative crosslinking which prevails initially. At longer UV exposures, chain-scission becomes the dominant reaction, thus resulting in the sharp decrease of the ultimate elongation after 4 days of photooxidation.

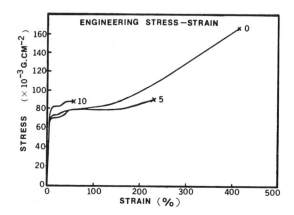

*Figure 1.    Engineering stress–strain curves as a function of time of UV exposure (numbers next to each curve represent days of exposure at 37°C)*

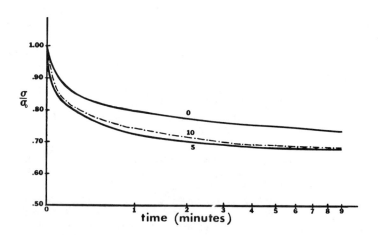

*Figure 2.    Stress–relaxation as a function of time of UV exposure (numbers on each curve represent days of exposure at 37°C)*

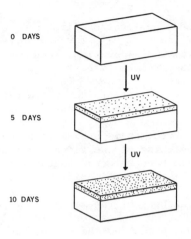

0  DAYS

UV

5  DAYS

UV

10 DAYS

*Figure 3.    Model of surface-layer oxidation*

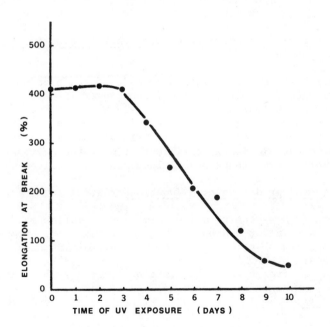

*Figure 4.    Ultimate elongation as a function of time of UV exposure*

## Effect of Strain on Photodegradation

a) <u>Fourier Transform infrared spectra</u>:  FTIR spectra of the following samples are shown in Figure 5:

     a)   PE film, non-prestrained, non-photooxidized,

     b)   PE film, exposed to UV light at 37°C for 120 hours,

     c)   PE film, prestrained to 200% elongation, then exposed to UV light at same temperature, for same period while kept stretched.

The c-spectrum shows more carboxylic content (ketonic absorbance at 1716 cm$^{-1}$), meaning that the prestrain has enhanced the photodegradation of the polymer.

It must be pointed out that a correction is needed to account for the change in thickness upon stretching of the films.  For that reason all carbonyl contents have been referred to the thickness of the non-strained film which has been photooxidized at the same temperature for the same period.  The 2840 cm$^{-1}$ band, which corresponds to the symmetric C-H stretching of the methylene groups (<u>14</u>) was chosen as reference for thickness correction (other bands can be used, but the choice was decided by the fact that the 2840 cm$^{-1}$ band changed in intensity as the strain varied, but its location and shape were not affected by the strain variation.)  Thus, all the photooxidation data are given on a relative scale.

b) <u>Strain-Enhanced Photodegradation as a Function of UV Exposure</u>:

Figure 6 shows the variation of photodegradation as a function of time of UV exposure for samples prestrained to different elongations as indicated by the numbers next to each curve.  The degradation rate increases as the strain increases from 0 to $\sim$ 300% and then decreases at higher strains.  This degradation rate dependence on strain is better illustrated if the degradation extent is plotted as a function of elongation for a given UV exposure.

c) <u>Photodegradation as a Function of Strain</u>:

Figure 7 shows the relative oxidation variation as a function of nominal strain for 5 days of UV exposure at 37°C.  From the data, it appears that the enhanced photodegradation occurs in three stages:

   <u>Stage I</u>:  from 0 to $\sim$ 120% strain.  Only a small increase of degradation rate is seen.

   <u>Stage II</u>: from 120% to $\sim$ 300% elongation.  A very pronounced increase of degradation rate is observed.

   <u>Stage III</u>:  300% to $\sim$ 500% elongation.  After an appreciable decrease, the degradation rate levels off.

An explanation of the presence of these different stages in

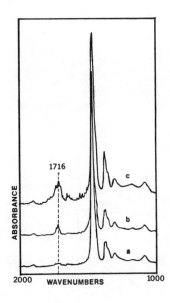

*Figure 5. FTIR: (a) nonoxidized, nonstrained sample; (b) nonstrained, oxidized (120 h of UV exposure at 37°C) sample; (c) prestrained (200%), then oxidized sample (same oxidation conditions as (b))*

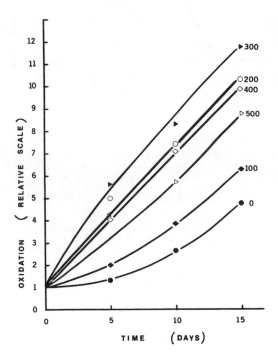

*Figure 6. Oxidation as a function of time of UV exposure for different prestrains (% prestrain indicated by numbers next to each curve)*

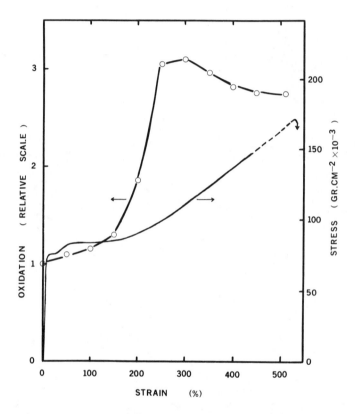

*Figure 7. Engineering stress (right ordinate) and oxidation (left ordinate) as a function of strain for a given UV exposure (5 days at 37°C)*

the enhancement process should not only take into account the photooxidation mechanism, but also the morphological changes induced upon stretching of the PE films. These morphological changes have been well investigated by Peterlin who proposed a model of plastic deformation of PE (15). This model, involving a deformation process of three stages, can be summarized as follows:

Stage I: Continuous deformation of the spherulitic structure before the neck;

Stage II: Discontinuous transformation in the neck, of the spherulitic into the fibrillar structure;

Stage III: Plastic deformation of the fibrillar structure after neck formation is complete.

The correlation between the stages of deformation and those of the enhanced photodegradation is clearly shown in Figure 7 where both nominal stress-strain curve (right ordinate) and photooxidation vs. elongation curve (left ordinate) are plotted. The close correspondence between the different stages of photooxidation and deformation allows us to make the following suggestions:

Stage I: 0 to $\sim$ 120% strain; the structure of the elongated polymer is still very close to that of the original material, and, therefore, no large difference is seen in the photodegradation rate.

Stage II: 120% to $\sim$ 300% strain; this stage corresponds more or less to the necking development region. In this stage drastic morphological changes occur via tilting, slippage and twisting of the lamellae, pulling of some chains out of the crystals, formation of microcracks and microvoids, all of which result in a highly disrupted structure. This disruption should greatly favor oxidation and this is seen in the very pronounced increase of carbonyl content in this stage.

Stage III: from $\sim$ 300% up to sample failure ($\lambda \sim 5.10$). The neck is more or less fully developed and a fibrillar structure is obtained. This structure is less susceptible to degradation because of its high degree of orientation and high crystallinity. This explains the drop and then levelling off of the carbonyl content in this latter stage.

In the previous experiment, the samples were kept strained during UV exposure and the taking of FTIR spectra. Thus, the enhancement of degradation may be attributed to a combination of both stress and strain effects. The strain effects, i.e., effects due to morphological changes induced by the deformation, are clearly shown in Stage II. During the necking development region, the applied load remains more or less constant while the elongation increases. Thus, the enhancement occurring in Stage II can be attributed mostly to strain effects. The stress effects, that is effects due to the applied load per se (or stored energy), are manifested by the fact that despite higher crystallinity-therefore less oxidation susceptibility-the highly drawn

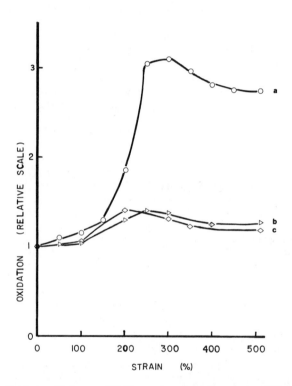

Figure 8.   Oxidation (5 days at 37°C) as a function of strain for different samples:
(a) prestrained with fixed-ends; (b) prestrained, then annealed with free-ends (for
one day at 60°C); (c) prestrained, then relaxed at room temperature with free-ends

polymers, $\lambda$ > 4, still show more degradation than the non-
strained polymer. This can be attributed to the higher reactivity toward oxidation of the stressed chemical bonds (16). It
was shown that under the action of mechanical stress, a homolytic
scission of macromolecular chains occurs and free radicals are
formed. The "mechanical" radicals will enhance the oxidation and
will initiate local fracture in stressed polymers (17).

## Effects of Annealing and Relaxation with Free-Ends

The next step was to see what happens to the enhanced photodegradation when the stress is relieved and the morphological
changes are reduced (free-end samples). The data shown in Figure
8-b were obtained using the samples that were first elongated,
then taken out of the stretcher and annealed with free-ends at
60°C for 24 hours before being photooxidized. The data in Figure
8-c were obtained using the samples which were first strained,
then taken out of the stretcher and allowed to relax at room
temperature for one week before photooxidation.

In both cases, the samples were exposed to UV light at 37°C
for 5 days. As it can be seen, the photodegradation still appears to have three stages corresponding to those observed in the
first case (Fig. 8-a), but there is a two-fold decrease in the
scale of the degradation. The overall effects of annealing and
relaxation are to allow the deformed structure to reorganize itself and partially return to the original relaxed state via elas-
tic and viscoelastic recovery. This recovery is not complete
even after a period of one week, and, on a molecular level, most
likely consists of small chain shifts in the crystalline lattice
and in lamellae rotation and slip. Thus, the effects of annealing and relaxation are to partially reverse the effects of
drawing (18) and this results in the reduction of the enhancement
of degradation. The fact that the necking development region
still shows a higher level of degradation suggests that in this
stage, the structure remains in a significantly disrupted state
and, therefore, still shows more susceptibility to degradation
even after annealing and relaxation.

## Conclusions

The photodegradation of low density polyethylene films is
greatly enhanced by uniaxial elongation, and the enhancement process is closely related to the morphological changes induced upon
drawing of the polymer films. The necking development region
shows more degradation due to its highly disrupted structure.
The enhancement can be reduced by annealing or relaxation, and
stress (applied load, or stored energy) appears to be the dominant factor of enhancement. The mechanical tests confirmed that
the oxidation is concentrated in the surface layers.

## Acknowledgement

The Fellowship support of SONATRACH (National Oil and Gas Company of Algeria) is gratefully acknowledged.

## Literature Cited

1. Jellinek, H.H.G., "Degradation of Vinyl Polymers", Academic Press, New York, 1955.
2. Raff, R.A.V.; Doak, K. W., Eds. "Crystalline Olefin Polymers", Part II, Interscience, New York, 1964, Chap. 8.
3. Pinner, S. H., Ed. "Weathering and Degradation of Plastics", Gordon and Breach, London, 1966.
4. Kamal, M. R., Ed. "Weatherability of Plastic Materials", Appl. Polym. Symp., 4, Interscience, New York, 1967.
5. Neiman, M. B., Ed. "Aging and Stabilization of Polymers", Consultant's Bureau, New York, 1965.
6. Scott, G., "Atmospheric Oxidation and Antioxidants", Elsevier, New York 1965.
7. Ershov, Yu. A.; Kuzina, S. I.; Neiman, M. B., Russ. Chem. Rev., 1969, 38, 147.
8. Hawkins, W. L., "Oxidative Degradation of High Polymers", in Oxidation and Combustion Reviews, Tipper, C.F.H., Ed., Vol. I, Elsevier, New York, 1965.
9. Howard, K. W., "The Effects of Weathering on the Engineering Behavior of Plastic Films", Ph.D. Thesis, University of California, Davis, 1976.
10. Kaufman, F. S., Jr., "A New Technique for Evaluating Outdoor Weathering Properties of High Density Polyethylene", in ref. 4.
11. Trozzolo, A. M., "Stabilization Against Oxidative Photodegradation", in "Polymer Stabilization", Hawkins, W. L., Ed., Wiley-Interscience, New York, 1972.
12. McKellar, J. F.; Allen, N. S., "Photochemistry of Man-Made Polymers", Applied Science Publishers, London, 1979.
13. Ranby, B.; Rabek, J. F., "Photodegradation, Photooxidation and Photostabilization of Polymers", John Wiley & Sons, New York, 1975.
14. Silverstein, R. M.; Bassler, G. C.; Morrill, T. C., "Spectrometric Identification of Organic Compounds", 3rd Ed., John Wiley & Sons, New York, 1974.
15. Peterlin, A., J. Mat. Sci., 1971, 6, 490.
16. Zhurkov, S. N.; Kosukov, V. E., J. Polym. Sci., Polym. Phys. Ed., 1974, 12, 385.
17. Pratt, P. L., "Fracture", Ed., Chapman and Hall, London, 1969, p. 531.
18. Peterlin, A., Makromol. Chem., Suppl., 1979, 3, 215.

RECEIVED October 17, 1980.

# Halogenated Polymethacrylates for X-Ray Lithography

A. ERANIAN, A. COUTTET, E. DATAMANTI, and J. C. DUBOIS

Thomson-CSF Laboratoire Central de Recherches, Domaine de Corbeville, 91401 Orsay, France

A large increase in the functional complexity and density of integrated circuit devices has occurred during the last ten years. The greater this density has been, the more important the necessity of higher resolution has become. Photolithography has been used extensively to replicate patterns of linewidth greater than 1 μm but its main restriction arises from diffraction effects which limit further resolution. Electron beam lithography provides finer patterns (0.3 μm linewidth) but the writing speed depends on the complexity of the patterns to be drawn. Moreover, the resolution is further limited by scattering effects due to the high electron energies generally used. X-ray lithography was proposed by Spears and Smith (1) in 1972 as a suitable technique for manufacturing very fine patterns of submicron size with both high speed and improved resolution since it avoids the essential shortcomings of the two previous techniques. First, diffraction is negligible because the wavelength used (4-50 Å) is shorter than that of ultraviolet light. Secondly, scattering is minimized because X-rays eject photoelectrons whose energies (less than 3 keV) are considerably lower than those used in electron beam lithography.

The requirements of speed and resolution not only determine the type of radiation to be used but also the nature of the resist to be exposed. Therefore, two resist characteristics appear to be particularly important :
- The sensitivity σ expressing its ability to have its structure modified as a result of exposure (more sensitive resists require lower exposure doses).
- The contrast γ defining its ability to provide steep pattern profiles.

It is well known that resists likely to be used in any of these lithographic methods are classified into two groups according to their behavior under irradiation :
- negative resists which crosslink and become insoluble,
- positive resists which undergo chain scission leading to more soluble fragments.

0097-6156/81/0151-0275$05.00/0

Negative resists generally exhibit high sensitivity but low contrast. For instance, in our laboratory, polymers containing thiirane groups $\overset{C}{\diagup}\overset{}{S}\overset{C}{\diagdown}$ were found to be extremely sensitive ($\sigma = 6 \times 10^{-7}$ C/cm$^2$) to electron beam irradiation at a 20 kV accelerating voltage but to have a low contrast $\gamma$ very close to unity (2). Positive resists, on the other hand, exhibit higher contrast but low sensitivity. An outstanding example is poly(methyl methacrylate) where $\sigma = 10^{-4}$ C/cm$^2$ and $\gamma > 2$ for electron beam irradiation of PMMA Elvacite 2041 with a 20 kV accelerating voltage.

One method for obtaining a high masking speed and resolution with X-ray lithography is use of highly sensitive positive resists. This paper reports some investigations on such sensitive positive X-ray resists.

## Parameters affecting the sensitivity of positive X-ray resists

It has been noted previously that radiation exposure of a positive resist causes chain scission in its structure. As a result, its molecular weight decreases. According to Ku and Scala (3), the decrease in the number average molecular weight resulting from such irradiation is given by the expression :

$$\bar{M}_n' = \frac{\bar{M}_n}{1 + p_s \bar{M}_n} \qquad (1)$$

where $\bar{M}_n$ is the number average molecular weight before radiation exposure. $\bar{M}_n'$ is the number average molecular weight after exposure and $p_s$ is the probability of scission, which depends both on the absorbed energy $\sigma_{abs}$ and on molecular parameters that can be expressed as :

$$p_s = \frac{\sigma_{abs} \ (\text{eV/cm}^3) \ G(s)}{100} \ \frac{M_o}{\rho \ \mathscr{N}_A} \qquad (2)$$

where $G(s)$ is the number of scission events occurring in the resist per 100 eV absorbed, $M_o$ is the monomer molecular weight, $\rho$ is the density of the resist, $\mathscr{N}_A$ is Avogadro's number. Substituting (2) into (1) leads to :

$$\sigma_{abs} \ (\text{eV/cm}^3) = \frac{100 \ \mathscr{N}_A \ \rho}{M_o \ G(s) \ \bar{M}_n} \left( \frac{\bar{M}_n}{\bar{M}_n'} - 1 \right) \qquad (3)$$

The ratio $\bar{M}_n/\bar{M}_n'$ can be expressed as a ratio of solubility rates before and after irradiation according to the relationship given by Greeneich (4).

$$R_{ni} = R_o + \beta \, (\bar{M}_{ni})^{-\alpha} \tag{4}$$

where $R_{ni}$ is the solubility rate in the developer mixture of a fragment of average molecular weight $\bar{M}_{ni}$, $\alpha$ and $\beta$ are constants characteristic of the developer mixture at a given temperature and $R_o$ is the background solubility rate (which can be assumed to be negligible for an optimum developer mixture). The exposure time $T_{exp}$ needed in X-ray lithography depends on the absorbed X-ray dose $\sigma_{abs}$ according to the following expression (5) :

$$T_{exp} = \frac{\sigma_{abs}}{\Phi \left(\dfrac{\mu}{\rho}\right)\rho} \tag{5}$$

where $\Phi$ is the incident X-ray flux, $(\mu/\rho)$ is the photoelectric mass absorption coefficient of the resist at the wavelength used and $\rho$ is the density of the resist.

Combining (3), (4), (5) leads to the expression of the incident dose required for X-ray exposure (i.e. the sensitivity) expressed as $\Phi \times T_{exp}$ :

$$\Phi \times T_{exp} = \frac{100 \; \mathscr{N}_A}{M_o} \times \frac{1}{\left(\dfrac{\mu}{\rho}\right) G(s)} \times \frac{\left[\left(\dfrac{R_n{}'}{R_n}\right)^{\frac{1}{\alpha}} - 1\right]}{\bar{M}_n} \tag{6}$$

Thus it can be seen that X-ray sensitivity for a positive resist depends on :
- the mass absorption coefficient of the resist at the irradiation wavelength,
- the number of scission events per 100 eV absorbed,
- the original number average molecular weight and,
- a solubility factor involving the solubility rates before and after irradiation.

This work has been mostly concerned with increasing the $(\mu/\rho)$ of poly(methacrylates) in order to improve sensitivity (i.e. to lower $\Phi \times T_{exp}$), although designing special structures will probably also affect the other parameters.

X-ray absorption and chemical design considerations

The need for a high value of $(\mu/\rho)$ partly explains why soft X-rays (4-50 Å) are more suitable than hard X-rays for exposing the resists, because $(\mu/\rho)$ increases very strongly with wavelength between absorption edges according to the Bragg-Pierce law :

$$\frac{\mu}{\rho} = C \; \frac{\mathscr{N}_A}{A} \; Z^4 \; \lambda^3 \tag{7}$$

where C is a constant for each curve segment, and A and Z are the
atomic weight and atomic number, respectively. Moreover, it is
obvious from (7) that high Z atoms provide a substantial absorption. Thus, incorporating heavy atoms in resist structures would
be very interesting since the mass absorption coefficient is an
additive quantity. For a compound of molecular weight M containing
$\alpha_i$ atoms i of atomic weight $A_i$, the whole mass absorption coefficient can be expressed as :

$$\frac{\mu}{\rho} = \sum_i \left(\frac{\mu}{\rho}\right)_i \times \frac{A_i}{M} \times \alpha_i \qquad (8)$$

Due to the existence of absorption edges, there are some wavelength restrictions which must be considered in order to take
full advantage of the absorption of a given atom. As shown in
Figures 1a and 1b where the mass absorption coefficients of the
elements are plotted versus their atomic number at two usable X-
ray wavelengths (8.34 Å corresponding to the $K\alpha_{1,2}$ emission line
of aluminum and 13.34 Å corresponding to the $L\alpha_{1,2}$ emission line
of copper), it turns out that transition metals in general would
be suitable if only absorption were considered.
    The incorporation of such metals into polymer structures,
particularly into poly(methacrylates) or doping these polymers
with organometallic compounds often affects their solubility
properties. Thus, the amount to be used is limited. Therefore, we
chose to incorporate halogen atoms such as fluorine, chlorine and
bromine which are less absorbing than metallic elements but
permit suitable solubility. In Table I are listed the photoelectric mass absorption coefficients of halogen atoms at the previously quoted wavelengths, compared to those of carbon, oxygen
and hydrogen which are found in many common polymers.

<u>Structure, preparation and characterization of the polymers</u>

    a) <u>Structure</u>   The polymers we have investigated are poly
(halogenoalkyl methacrylates) of the following typical structure :

$$\left(\begin{array}{c} CH_2 - \overset{\displaystyle CH_3}{\underset{\displaystyle C}{\overset{|}{\underset{|}{C}}}} \\ \end{array}\right)_n$$

$$\begin{array}{c} O \diagup \diagdown O \\ | \\ CH_2 \\ | \\ CH_{(3-a)} \ X_a \end{array}$$

with a ≤ 3 and X is fluorine, chlorine or bromine. Specifically,
we have synthesized the following polymers :

X = F, a = 1 : poly(2-fluoroethyl methacrylate) PFEMA
X = Cl, a = 1 : poly(2-chloroethyl methacrylate) PClEMA
X = Br, a = 1 : poly(2-bromoethyl methacrylate) PBrEMA
to study the influence of the nature of the halogen atom. In addition we have studied X = F, a = 3 : poly(2-,2-,2-trifluoroethyl methacrylate) PF$_3$EMA to analyze the effect of increasing the number of halogen atoms.

The photoelectric mass absorption coefficients of these polymers and of poly(methyl methacrylate) PMMA at 8.34 Å and 13.34 Å are shown in Table II. These coefficients were calculated using relation (8) and the data listed in Table I.

b) <u>Preparation and characterization</u>   Monomers were obtained either by esterification of methacrylic acid by the corresponding halogenated alcohols or by acylation of those alcohols by methacryloyl chloride. The structures were evaluated using elemental analysis and infrared spectrometry.

All of the polymers were prepared by free-radical procedures at 80°C for 48 hours using benzene as solvent, benzoyl peroxide as catalyst and a monomer/solvent weight ratio of 0.64. They were isolated by precipitation with methanol and dried in a vacuum. Elemental analysis and infrared spectrometry were used to determine the structures. For each polymer structure, several samples of different molecular weight were synthesized by varying the catalyst concentration. They were characterized by intrinsic viscosity measurements as shown in Table III which also lists the preparative conditions. Attempts were made to get higher molecular weights by reducing catalyst concentration further. In the case of PFEMA they lead to insoluble unusable polymers. In the case of PBrEMA, no molecular weight increase was observed, probably due to extensive chain transfer.

X-ray exposure devices

Both characteristic X-ray line and continuous spectra were used to evaluate the performances of the resists. To determine exposure parameters (i.e. sensitivity and contrast) irradiations were carried out in this study using the aluminum Kα$_{1,2}$ emission line at 8.34 Å generated by means of a modified Vacuum Generators Limited model EG-2 electron beam evaporation gun. The resist samples were exposed through a mask (A) consisting of a range of aluminum foils of different thicknesses supported on an absorbing nickel frame in order to vary the X-ray flux.

To test the feasibility of obtaining submicron size patterns in the resist films, an exposure source was used which consisted of the X-ray continuous spectrum produced by synchrotron radiation from the 540 MeV storage ring of the University of Orsay (ACO) since synchrotron radiation had been shown previously (7,8) to be a suitable source for providing very high resolution due to the small divergence of the beam. The maximum output flux of ACO

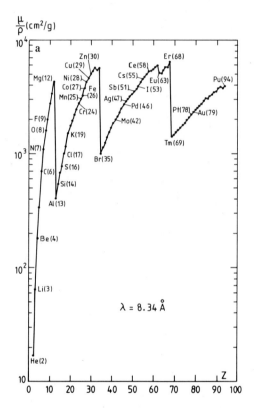

*Figure 1.   X-ray mass absorption coefficient vs. atomic number at (a) 8.34 Å (alu-
minum $K\alpha_{1,2}$ line) and (b—facing page) 13.34 Å (copper $L\alpha_{1,2}$ line)*

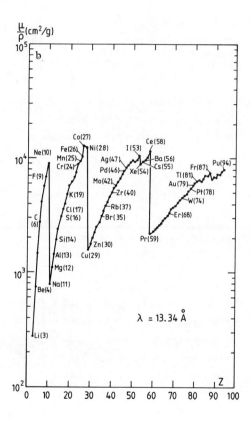

**Table I.  Photoelectric Mass Absorption Coefficients of Halogen Atoms and of Carbon, Oxygen, Hydrogen at 8.34 Å (Al $K\alpha_{1,2}$ emission line) and 13.34 Å (Cu $L\alpha_{1,2}$ emission line) (6)**

| $\frac{\mu}{\rho}$ (cm$^2$/g) | H | C | O | F | Cl | Br |
|---|---|---|---|---|---|---|
| 8.34 Å (Al : $K\alpha_{1,2}$) | 1.8 | 718.4 | 1597 | 2037 | 1023 | 1021 |
| 13.34 Å (Cu : $L\alpha_{1,2}$) | 8.9 | 2714 | 5601 | 6941 | 3596 | 3101 |

Advances in X-Ray Analysis

**Table II.  Photoelectric Mass Absorption Coefficients of the Halogenated Polymethacrylates and of PMMA**

| Polymer | Monomer formula | $(\mu/\rho)$ (cm$^2$/g) | |
|---|---|---|---|
| | | $\lambda = 8.34$ Å | $\lambda = 13.34$ Å |
| PMMA | $C_5 H_8 O_2$ | 941.5 | 3418.8 |
| PFEMA | $C_6 H_9 O_2 F$ | 1071.5 | 3835.1 |
| PClEMA | $C_6 H_9 O_2 Cl$ | 936.5 | 3381 |
| PBrEMA | $C_6 H_9 O_2 Br$ | 955.6 | 3225.6 |
| PF$_3$EMA | $C_6 H_7 O_2 F_3$ | 1302.5 | 4582.8 |

occurs around 14 Å for the 540 MeV energy. The mask (B) used in
those experiments consisted of 0.1 µm thick absorbing gold patterns evaporated on a membrane of 0.5 µm thick silicon nitride
supported on a silicon frame.

Evaluation of the polymers as X-ray resists

    The polymers were dissolved in methylisobutylketone (MIBK)
and spin-coated on oxidized silicon wafers (1100 Å thick $SiO_2$
layers) to form 5000 Å thick films. After a prebaking to improve
adhesion to the substrate, the resist samples were irradiated
through the mask (A) using the Al $K\alpha_{1,2}$ emission line at 8.34 Å
as X-ray source. The electron beam gun was operated at a 300 W
power and the source to sample distance was 4.9 cm. Taking into
account the absorption of the aluminum foil mask, the different X-
ray fluxes available on the sample were calculated from the
relation given by ($\underline{9}$) :

$$\Phi = \frac{I \ \varepsilon \ h\nu}{D^2 \ e}$$ (9)

where I is the electron beam current, $\varepsilon$ is the quantum efficiency
of the target, $h\nu$ is the photon energy, D is the source to
sample distance and e is the electron charge. The exposed samples
were then developed in a solvent/non-solvent mixture at 30°C for
1 min. The solvent was either methylethylketone (MEK) or methyl-
isobutylketone (MIBK) whereas isopropyl alcohol (IPA) was the
non-solvent in each case. The solvent/non-solvent ratio was
adjusted so that the unexposed areas on samples remained unaffected during development. The sensitivity was determined from
thickness measurements with a Sloan Co. Dektak Surface profile
measuring system. Plotting the normalized resist thickness (ratio
of remaining thickness $e_r$ to original thickness $e_0$) versus the
incident X-ray dose ($\Phi$ x $T_{exp}$) provides the sensitivity, as the
dose corresponding to the total solubility of the exposed areas
(normalized thickness = 0). The contrast was evaluated as the
slope of the X-ray dose-thickness curve interpolated between the
zero and 0.7 values of the normalized thickness.
    Table IV compares the X-ray exposure characteristics (at
8.34 Å, Al $K\alpha_{1,2}$ emission line) of the halogenated resists and of
PMMA Elvacite 2041. It can be seen that poly(2-chloroethyl methacrylates) and poly(2-bromoethyl methacrylates) exhibit a low
sensitivity unlike poly(2-fluoroethyl methacrylates) and poly(2-,
2-,2-trifluoroethyl methacrylates) which are more sensitive than
PMMA as shown in Figures 2a, 2b, 2c, 2d where the dose-thickness
curves of these resists are plotted. The low sensitivity of the
PClEMA and PBrEMA samples may be explained by some competing
crosslinking reactions which could occur during exposure as a
result of C-Cl and C-Br homolytic bond scissions as noted by Tada

**Table III.  Conditions for Polymer Preparation and Sample Characterization**[a]

| Polymer name | sample | % catalyst concentration | % yield | intrinsic viscosity $\eta$ (dl/g) at 30°C |
|---|---|---|---|---|
| PFEMA | 1 | 0.1 | 96.2 | 0.46* |
| PFEMA | 2 | 0.05 | 85.2 | 0.88* |
| PClEMA | 1 | 0.24 | 86.6 | 0.20** |
| PClEMA | 2 | 0.01 | 47.5 | 1.11** |
| PBrEMA | 1 | 0.1 | 75.8 | 0.26** |
| PBrEMA | 2 | 0.04 | 45.5 | 0.35** |
| PF$_3$EMA | 1 | 0.2 | 92.1 | 0.21* |
| PF$_3$EMA | 2 | 0.1 | 87.1 | 0.31* |
| PF$_3$EMA | 3 | 0.05 | 85.7 | 0.45* |

[a] (*) from measurements in acetone; (**) from measurements in benzene; (***) for comparison $\eta$ (PMMA Elvacite 2041) = 1.11 dl/g (in benzene at 30°C).

**Table IV.  X-ray Exposure Characteristics (at 8.34 Å : Al K$\alpha_{1,2}$ emission line) of the Poly(halogenoalkyl methacrylates) and of PMMA Elvacite 2041**[a]

| Polymer name | sample | intrinsic viscosity at 30°C (dl/g) | Developing mixture *** (1 min at 30°C) | Dose $\sigma_0$ (mJ/cm$^2$) | Contrast value |
|---|---|---|---|---|---|
| PMMA | elvacite 2041 | 1.11** | MEK / IPA 1:4 | 2180 | 2.4 |
| PFEMA | 1 | 0.46* | MIBK/IPA 3:1 | 730 | 3.4 |
| PFEMA | 2 | 0.88* | MIBK/IPA 3:1 | 730 | 4.2 |
| PClEMA | 1 | 0.20** | MEK / IPA 1:4 | High | – |
| PClEMA | 2 | 1.11** | MEK / IPA 1:4 | High | – |
| PBrEMA | 1 | 0.26** | MIBK/IPA 1:1 | High | – |
| PBrEMA | 2 | 0.35** | MIBK/IPA 1:1 | High | – |
| PF$_3$EMA | 1 | 0.21* | MIBK/IPA 1:4.5 | 1255 | 1.8 |
| PF$_3$EMA | 2 | 0.31* | MIBK/IPA 1:3.5 | 730 | 2.4 |
| PF$_3$EMA | 3 | 0.45* | MIBK/IPA 1:3.5 | 730 | 3 |

[a] (*) in acetone; (**) in benzene; (***) solvent/nonsolvent ratios are volume ratios.

(10). Thus, the number of scission events per 100 eV absorbed would be decreased. The sensitivity of the best samples of PFEMA and PF$_3$EMA resists was found to be the same : 730 mJ/cm$^2$, a factor of three greater than that of PMMA. For both resists the contrast seems to increase with the intrinsic viscosity, i.e. with the molecular weight, but it is better for PFEMA resist. As expected from absorption considerations (see ($\mu/\rho$) values in Table II) the sensitivity of PF$_3$EMA should have been higher than that of PFEMA since the intrinsic viscosity (i.e. the molecular weight in rough approximation) was the same. The fact that this was not the case could be due to different solubility properties. Besides, referring to PFEMA sample 1 and PF$_3$EMA sample 3 ($\eta$ = 0.45 dl/g for both resists), the former was developed by a 3 : 1 MIBK/IPA ratio whereas development of the later needed a 1 : 3.5 ratio.

Further experiments concerning the development of these fluorinated resists were conducted. As shown in Figures 3a and 3b, the sensitivity is strongly dependent on the developer and indeed MIBK/IPA was found to be the most suitable mixture for both PFEMA and PF$_3$EMA. Moreover, as far as masking properties were concerned, when development was performed at a temperature lower than 30°C for a time longer than 1 min, higher resolution was obtained. Therefore, development conditions were adopted at 20°C for PFEMA resist : MIBK/IPA 4 : 1 ratio for 150 sec. These new conditions were used when resolution had to be tested. Figures 4a and 4b reproduce scanning electron micrographs of patterns with 1 µm to 0.3 µm wide lines drawn in a 0.8 µm thick film of PFEMA exposed through the mask (B) to synchrotron radiation (X-ray continuous spectrum) from the 540 MeV electron synchrotron ACO in Orsay. The exposure time was 2.5 times shorter than that required for PMMA.

Initial studies of the dry-etch resistance of these resists showed that PFEMA had an etching rate in a CCl$_4$ plasma (p = 0.2 torr, P = 600 W) about three times lower (i.e. 380 Å/min) than that of aluminum (i.e. 1200 Å/min).

Conclusions

The molecular parameters affecting the sensitivity of positive X-ray resists have been outlined. Incorporating halogen atoms into the resists has been shown as one way to increase sensitivity. Among the halogenated poly(methacrylates), two have been found to be very interesting : poly(2-,2-,2-trifluoroethyl methacrylate) PF$_3$EMA and poly(2-fluoroethyl methacrylate) PFEMA because their sensitivity was three times higher than that of PMMA at 8.34 Å and because their contrast was good, especially that of PFEMA. Using PFEMA resist, 0.3 µm wide lines have been obtained in a 0.8 µm thick film exposed to synchrotron radiation. Moreover, it has been shown to be a suitable resist for aluminum plasma etching due to its resistance to a CCl$_4$ plasma.

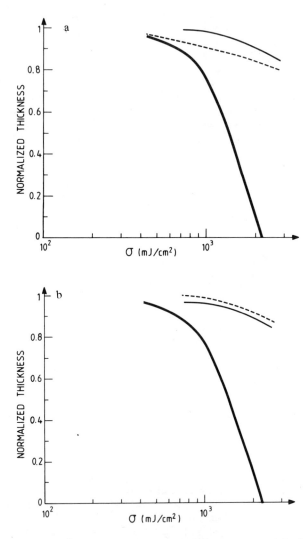

*Figure 2.   X-ray dose–thickness curves of (a) PClEMA (λ = 8.34 Å: Al Kα₁,₂ emission line): (———) η = 1.11 dl/g (Sample 2); (– – – –) η = 0.2 dl/g (Sample 1); (▬▬) (PMMA Elvacite 2041) and (b) PBrEMA (λ = 8.34 Å: Al Kα₁,₂ emission line): (– – – –) η = 0.35 dl/g (Sample 2); (———) η = 0.26 dl/g (Sample 1); (▬▬) (PMMA Elvacite 2041).*

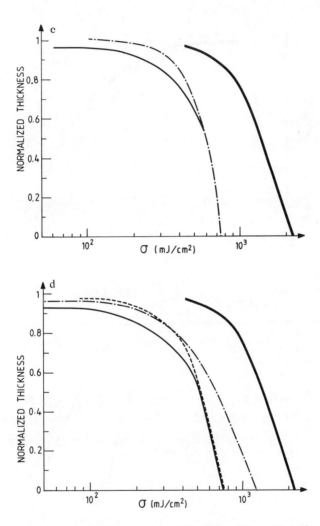

*Figure 2.   X-ray dose–thickness curves of (c) PFEMA ($\lambda = 8.34$ Å: Al $K\alpha_{1,2}$ emission line): (— · —) $\eta = 0.88$ dl/g (Sample 2); (————) $\eta = 0.46$ dl/g (Sample 1); (PMMA Elvacite 2041) and (d) PF$_3$EMA ($\lambda = 8.34$ Å: Al $K\alpha_{1,2}$ emission line): (— — —) $\eta = 0.45$ dl/g (Sample 3); (————) $\eta = 0.31$ dl/g (Sample 2); (— · —) $\eta = 0.21$ dl/g (Sample 1); (▬▬▬) (PMMA Elvacite 2041).*

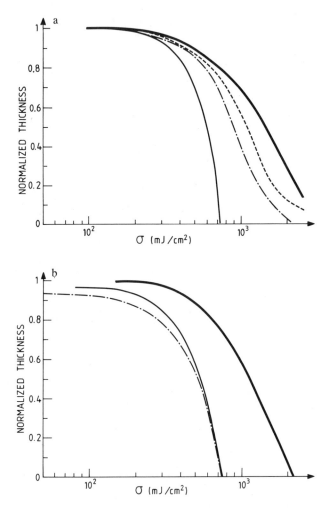

*Figure 3.    X-ray dose–thickness curves of (a) PFEMA Sample 2 (η = 0.88 dl/g)
for various developer combinations (λ = 8.34 Å: Al Kα₁,₂ emission line): (————)
MIBK/IPA  3:1; (——–—) ethyl acetate/IPA  1:1; (━━━━) chloroform/IPA  1:2;
(—–––) MEK/IPA 1:1.5; and (b) PF₃EMA Sample 3 (η = 0.45 dl/g) for various
developer combinations (λ = 8.34 Å: Al Kα₁,₂ emission line): (————) MIBK/IPA
1:3.5; (—– —) ethyl acetate/IPA 1:5; (━━━━) chloroform/IPA 1:2.*

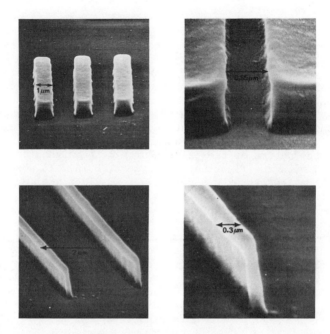

*Figure 4.   Scanning electron micrographs of patterns in a 0.8-μm PFEMA film exposed to synchrotron radiation from the French electron synchrotron ACO in Orsay (exposure time 2.5 times shorter than that required for PMMA) and developed in a MIBK/IPA 4:1 mixture at 20°C for 150 s*

## Acknowledgments

The authors would like to thank Dr. B. Fay for resist exposures by ACO synchrotron radiation, Dr. P. Parrens from LETI (Grenoble) for measurement of plasma etching rates and the "Direction des Recherches, Etudes et Techniques" (DRET) which partly sponsored this work.

## Abstract

The sensitivity of positive resists for X-ray lithography depends on molecular parameters such as absorbed X-ray energy. Thus, sensitivity enhancement can be obtained by incorporating heavy atoms. The need to keep solubility properties lead us to choose poly(halogenoalkyl methacrylates) for evaluating them as X-ray sensitive resists. Thus, both mono- and poly-halogenated poly(methacrylates) have been prepared and their behavior under soft X-ray irradiation has been studied. The sensitivity of poly(2-monohalogenoethyl methacrylates) with different halogen atoms, i.e. fluorine, chlorine and bromine, has been found to depend on the nature of the halogen atom ; the highest sensitivity is obtained with the fluorinated polymer (PFEMA). It works as a positive X-ray resist with a dose 2.5 to 3 times smaller than that required for PMMA Elvacite 2041. Masking patterns with 0.3 μm linewidth  have been drawn in a 0.8 μm thick film of PFEMA exposed to continuous X-ray radiation from the French electron synchrotron ACO in Orsay. Moreover, PFEMA resist appears to have a good resistance to $CCl_4$ plasma etching and can therefore be used for aluminum etching. Incorporating more fluorine atoms in the same length ester group of poly(methacrylates) in order to increase X-ray absorption also affects other properties such as solubility, thus explaining that poly(2-,2-,2-trifluoroethyl methacrylate) $PF_3EMA$ is not more sensitive than PFEMA.

## Literature cited

1.  Spears D.L., Smith H.I.
    Electron. Letters, 1972, **8**, 102.
2.  Gazard M., Dubois J.C., Eranian A.
    Regional Technical Conference, Soc. of Plastics Engineers "Photopolymers, Principles, Process and Materials", Nevele, New-York (October 1976).
3.  Ku H.Y., Scala L.C.
    J. Electrochem. Soc., 1969, **116** (7), 980.
4.  Greeneich J.S.
    J. Electrochem. Soc., 1974, **121** (12), 1669.
5.  Spears D.L, Smith H.I.
    Solid State Technology, 1972, **15**, 21.
6.  Henke B.L., Ebisu E.S.
    Advances in X-ray analysis, 1974, **17**, 150.

7.  Fay B., Trotel J.
    Appl. Phys. Lett., 1976, 29 (6), 370.
8.  Spiller E., Eastman D.E., Feder R., Grobman W.D., Gudat W.,
    Topalian J.
    J. Appl. Phys., 1976, 47 (12), 5450.
9.  Lenzo P.V., Spencer E.G.
    Appl. Phys. Lett., 1974, 24 (6), 289.
10. Tada T.
    J. Electrochem. Soc., 1979, 126 (9), 1635.

RECEIVED October 17, 1980.

# INDEX

# INDEX